# MOLECULAR AND CRYSTAL STRUCTURE MODELS

is Senior Lecturer in
eld College, University
ed the College initially
niversity of Aberdeen,
Assistant in Chemistry.
er B.Sc. and Ph.D. in
University of Notting-
0 respectively.

areer began in 1948,
Research Association,
facturers as a labora-
1952 to 1954 she was
t Beckenham County
Girls, Kent, simul-
part-time at Croydon
echnics for her uni-
ications.

esteemed in the field
stal structure models,
demonstrated on
cations at the Univer-
h, Nottingham, East
sity of London Insti-
d mounted full-scale
exhibitions at Westfield College and at the University of Sheffield for The Chemical Society Autumn Meetings, 1976.

James Gillison's cover design shows a *guanine-cystosine base pair* model

# Molecular and Crystal Structure Models

Anne Walton,
Senior Lecturer in Chemistry
Westfield College
University of London

ELLIS HORWOOD LIMITED
Publishers Chichester

Halsted Press, a division of
JOHN WILEY & SONS
New York · Chichester · Brisbane · Toronto

*The publisher's colophon is reproduced from James Gillison's drawing of the ancient Market Cross, Chichester*

First published in 1978 by

**ELLIS HORWOOD LIMITED**

Market Cross House, 1 Cooper Street, Chichester, Sussex, England

**Distributors:**

*Australia, New Zealand, South-east Asia:*
Jacaranda-Wiley Ltd., Jacaranda Press,
JOHN WILEY & SONS INC.,
G.P.O. Box 859, Brisbane, Queensland 40001, Australia.

*Canada:*
JOHN WILEY & SONS CANADA LIMITED
22 Worcester Road, Rexdale, Ontario, Canada.

*Europe, Africa:*
JOHN WILEY & SONS LIMITED
Baffins Lane, Chichester, Sussex, England.

*North and South America and the rest of the world:*
HALSTED PRESS, a division of
JOHN WILEY & SONS
605 Third Avenue, New York, N.Y. 10016, U.S.A.



**British Library Cataloguing in Publication Data**

Walton, Anne
Molecular and crystal structure models.
1. Molecules - models 2. Crystals – models
I. Title
539'.12'0228 QD480 78-40227
ISBN 0-85312-083-8 (Ellis Horwood)
ISBN 0-470-26356-3 (Halsted)



Typeset in Press Roman by Coll House Press, Chichester, Sussex

Printed in Great Britain by Butler and Tanner Ltd., Frome.

# Table of Contents

# Author's Preface

Since beginning teaching in 1960 I have used models more and more extensively as aids in both teaching and research. This trend reflects in part my personal professional interests in structural chemistry, and in part the increased emphasis which is now placed on structure in the teaching of chemistry. Because an understanding of structural chemistry is greatly facilitated by the use of models (see Chapter1), I felt that a book such as this might be opportune. In it I attempt both to survey most of the manufactured models, and also to give some hints on homemade ones, together with references.

At first my intention was to include information on all models currently available. However, as the writing progressed it became apparent that a lack of balance in the book as a whole would result and that it would be preferable to omit some examples. Similarly, some published work on homemade models has not been included. The choice, necessarily, has been idiosyncratic and to those whose own work or whose favourite model has been omitted, or who feel that the treatment has been inadequate, I tender my apologies. I intend no offence.

It is also possible that I have excluded some examples through ignorance of their existence. If such is the case, I should be most interested to hear about them and should be delighted if their creators would contact me.

Manufacturers tend to use trivial names for chemical compounds and units (such as inches) which would not be acceptable in a scientific paper. A reader interested in any of these models would presumably apply to a manufacturer for information and would receive a brochure or catalogue in which trivial names and non-S.I. units are used. In consequence I felt that applying systematic nomenclature to the common compounds to which these remarks apply, and converting to millimetres sphere sizes which are clearly multiples of an inch, would be mere pedantry. Hence I have used trivial names and old-fashioned units if, in the context, this seemed to be the most appropriate policy.

I have come to realise that, even before it reaches the publishers, the making of a book is complex, involving many people of diverse talents who contribute directly or indirectly in a variety of ways. Among my colleagues in the chemistry department at Westfield College I wish especially to thank Professor Bernard

Aylett, who encouraged me to write about models, Professor David Kirk, who has assembled many models of large organic molecules at my request and to acknowledge the assistance of the late Professor William Klyne who supplied me with much useful information. I should also like to thank Professor Malcolm Frazer and members of the Chemical Education Sector of the University of East Anglia, who helped me to finish this book by providing the necessary facilities during a period of study leave. Among these I include Mr. John Gray, who produced some photographs.

Next I wish to express my gratitude to Mrs. Bernice Benjamin of the Chemistry Department, Westfield College, who not only typed most of the manuscript but helped in numerous other ways; and also to Miss Julia Corbett, who assisted her.

My thanks are also due to Mr. Y. Sahota, head of the Reprographic Unit at Westfield College and to Mr. Keris Eyton-Williams, until recently also of the Reprographic Unit, who photographed many of the models.

Lastly, I am grateful to Miss Janet Chapman of the Spanish Department, Westfield College, who read the manuscript and suggested a number of improvements.

To all these colleagues and friends, this book is gratefully dedicated.

Chemistry Department,
Westfield College,
London

October 1977

# 1

# Introduction

## INTRODUCTION

'Up early and to my office; where Cooper came to me and begun his lecture upon the body of a ship – which my having of a modell in the office is of great use to me, and very pleasant and useful it is.'

S. Pepys; *Diary*, 30 July, 1662. [1]

## 1.1 WHAT IS A MODEL?

'If you can build a model of it, then you are well on the way to understanding the structure.' This is generally good advice, but what is meant by 'model' and 'understanding' in this context? The Oxford English Dictionary gives several definitions of 'model', the most relevant being, 'A representation in three dimensions of some projected or existing structure, or of some material object, artificial or natural, showing the proportions and arrangement of its component parts.' This covers almost all the categories of model described in this book because their chief function is to convey information concerning structure, relative dimensions and spatial distribution of the constituent parts in crystals or molecules.

A model is *not* a perfect replica, differing only in scale, of the 'real' thing. This is particularly true of molecular and crystal structure models as distinct from those of buildings, ships, aircraft and similar structures. All of these are scaled down, while massive scaling up (by factors of $10^8$ or more) is required for a molecular or crystal structure model. A major difficulty is that the atomic theory from which ideas of molecules and crystals derive is itself a model of the nature of matter, based on atoms, ions, electrons and molecules. This represents another problem in that our material models must of necessity represent a 'classical' rather than a quantum mechanical view of matter [2]. At best, therefore a model shows us only something of what a molecule is like. In other words, it emphasises particular features which accord with the facts about molecules as we know them [3]. Hence models can provide a link between observation and theory; experiment and imagination; familiar and unfamiliar [4], [5].

Whether we are dealing with theoretical or material models there is a case for starting with simple ones and working towards more complex examples [6]. The limitations of models must of course be recognised, and indicated when teaching, but their inadequacies should not deter us from using them. Langmuir has pointed out [7] that we cannot give absolute meaning to the set of abstractions which we use to describe natural phenomena. Therefore the model we choose must be one of several, all of which are inadequate. Molecular and crystal structure models, in common with many others, tend to fall into one of two groups. Either they are moderately successful compromises showing, somewhat vaguely, a number of structural features, or they are designed to display a particular aspect supremely well. Interestingly, 'less realistic' models are sometimes more effective than those in which primary function is obscured by too much attention to detail (see, for example, the butane models described by Petersen [8]). Thus one could argue that a mathematical model based on quantum theory is the best description of the properties of matter, but it is only completely successful when applied to the hydrogen atom [9].

Structural models are artefacts in which the bonding arrangements between atoms and ions believed to exist at an atomic level are simulated [10]. They are also analogues, [11] reflecting in their properties something of the fundamental nature of matter. Models as a means of communication have advantages over both the language of mathematics and that of words, having more complex logical operations than the latter and a wider vocabulary than the former [12]. Comprehension of what the model is intended to convey is often rapid although its inevitable inaccuracies detract from its utility. A model can be compared with an ideogram or a symbol in that its meaning is transmitted directly, not by means of some intermediary language.

Turning now specifically to molecular models, those aspects of molecular structure which can satisfactorily be portrayed by models have been classified by Ollis as constitution, configuration, conformation and chirality [13]. **Constitution** is concerned with which atoms are present in a molecule and how they are bonded together, while **configuration** specifies the spatial arrangements of atoms and bonds, *excluding* those which may arise by rotation about bonds. **Conformation,** on the other hand, involves *only* those spatial arrangements arising from rotation about bonds (for a given configuration). **Chirality** is a property of those forms which are not superimposable on their mirror images. Apart from these static considerations, Ollis also stresses the importance of molecular mobility, and mobile working models were demonstrated during his Evening Discourse at the Royal Institution.

Molecular models have been critically assessed by Mislow [14], who observes that bond angles are frequently inaccurate in models made from standard sets. He is also concerned at the lack of simulation of internuclear motion, in particular the stiffness of the bond angles in many models. Another problem is that manufactured models often work on the 'all-or-nothing' principle as regards rotation about bonds. In the model there is usually free rotation about single bonds and none about multiple bonds, a situation which does not necessarily obtain in the molecules themselves. Models are usually made to represent molecules at or near room temperature and no account is taken of the possibility of bonds either becoming free to rotate as the temperature is raised or restricted as the temperature is lowered.

The requirements for an ionic crystal structure model are different in that the chief interest lies in the immediate environment of each ion and how the ions are linked to form extended arrays. However, we again usually have a static model. Dynamic models have been designed (some of them are described in Chapter 8) but they are usually for a specific application and therefore limited in scope.

Most of the points touched on in the preceding three paragraphs are discussed in more detail in the Chapters dealing with particular model types.

There is evidence that models of chemical structures may be able to play an important part in the intellectual development of young people. It has been suggested that up to one half of the 'freshman nonscience majors' studying chemistry courses in Colleges in the USA have difficulty in understanding abstract concepts because they have not reached the 'formal operational' stage [15]. According to the Piaget classification, the concrete operational phase should give way to the formal phase at about the age of twelve. Intellectual development should be essentially complete by age fifteen. A great deal of the chemistry usually taught at the tertiary level requires formal thought for proper understanding, but students can sometimes 'get by' and even pass examinations if they are prepared to memorise a large proportion of the material given to them in class. This is an unfortunate state of affairs, but it can sometimes be remedied by careful teaching specifically planned to encourage the student to develop his latent faculties and achieve formal thought.

Models are an invaluable teaching aid in bridging the gap between the formal and concrete thought processes because the model *represents* an abstract idea but is itself 'concrete'. The model is a physical manifestation of a concept. Using it helps the student to imagine the concept and so helps to initiate him to formal thinking. A student at the concrete operational stage cannot visualise tetrahedra and octahedra as elements of structure without the aid of models [16]. I also suspect, although I have no proof as yet, that he also has difficulty even in relating two different types of model of the same structure where one consists of linked tetrahedra or octahedra and the other is, for example, a ball-and-spoke model. It is possible that in the student's mind the familiar ball-and-spoke model may have become synonymous with the structure itself and that the unfamiliar polyhedra are therefore unacceptable substitutes. Obviously pitfalls such as these must be overcome if the models are to be used successfully in aiding intellectual development.

## 1.2 CHEMICAL STRUCTURE MODELS

### 1.2.1 Classification

A possible classification scheme is given in Table 1.1 and the arrangement of the material in Chapters 2–8 is based on this. The 'dynamic' category is small compared to the great variety to be found under the 'static' heading. There is much overlap between types and in some cases classification into groups is arbitrary and difficult to rationalise. This is particularly true of the boundary between ball-and-spoke and skeletal models. Indeed, an alternative classification would be to call both these categories 'exploded', 'expanded' or 'open' models as opposed to the 'closed' space-filling ones. However, there is so much material on 'exploded' models that subdivision seemed to be necessary and it seemed sensible to retain the well-known terms. Linked polyhedra (Chapter 6) form a rather specialised category which finds application in advanced structural inorganic chemistry. Similarly distinct are models of orbitals, although they are often used in conjunction with other types of model. Some important 'subspecies' are omitted from the classification scheme as they do not form a sufficiently large group to warrant inclusion. An example is 'Biobits', a type used to construct models of macromolecules. This system is discussed in Chapter 6. Manufactured and homemade versions of each category can be purchased or made. Names and addresses of manufacturers and suppliers are listed in Appendix 1 and price-codes at the end of each Chapter. References to homemade models are numbered in the text and listed at the end of the appropriate Chapter.

**Table 1.1**

Classification of Molecular and Crystal Structure Models

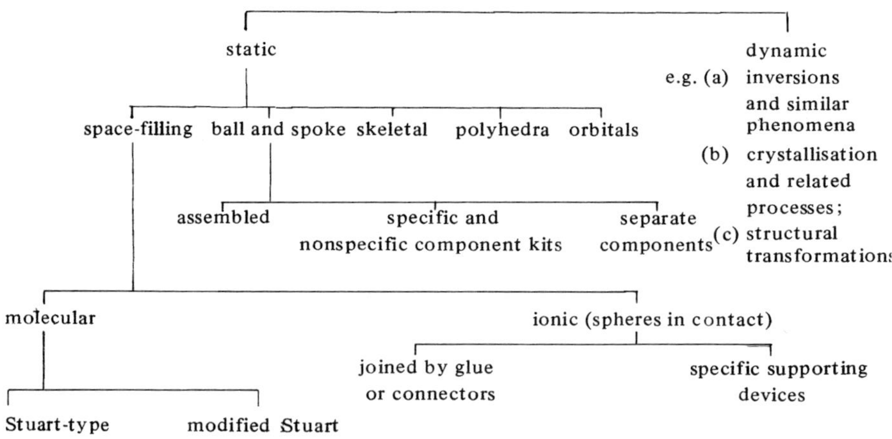

### 1.2.2 Colour Codes

In most molecular and crystal structure models a specific colour is used to represent a specific element. This practice is immensely valuable if one of the common colour codes is used because it enables one to 'read' an assembly very rapidly by translating the coloured models into atoms or ions. In fact interpretation is no more difficult, and in some ways much easier, than making chemical sense out of a formula. The letter code is simply replaced by coloured objects. It has been claimed that the colours chosen for the elements frequently used in models are those most suitable for people with defective colour vision. However, it seems probable that other considerations (perhaps unconscious ones) determined that carbon should be black, oxygen red (fire), chlorine green, sulphur yellow and so on. The most widely used code, certainly in Britain and America, is that of the Institute of Physics [17] (Table 1.2). The commonest variant occurs chiefly in the case of the halogens, which are all different colours instead of being deepening shades of green. Fluorine can be yellow, green or beige, chlorine is usually green, bromine brown and iodine violet. The Colour Group of the Physical Society recommends a colour code which is similar in most respects to that of the Institute of Physics, except that halogen is yellow.

**Table 1.2**
Institute of Physics Colour Code for Atom Models

| Element | C | H | O | N | F | Cl | Br | I | S | P | Si | Metals |
|---|---|---|---|---|---|---|---|---|---|---|---|---|
| Colour | black | cream or white | red | blue | yellowish green | light green | mid green | dark green | yellow | purple | black or grey | brown, silver or gold |

It would be of great advantage to model users if manufacturers could be persuaded to accept an international colour code. Of course preference for a particular colour code is largely a matter of habit, but it can be confusing when crucial elements such as carbon, oxygen, nitrogen and hydrogen are occasionally represented by different coloured models from those generally used. If the code is unknown (or forgotten!) it becomes necessary to work from the stereochemistry back to the model rather than the other way round.

### 1.2.3 Previously Published Surveys.

To my knowledge no book has so far been published which attempts to cover the whole range of model types as this one does. However, a number of surveys varying in length from a few hundred to about twenty thousand words have been published. Some of these are listed in Table 1.3. A number of books on models have been published, but each is concerned with a limited number of model types. References to these will be found in the appropriate Chapters.

**Table 1.3**
Some Surveys of Model Types

| Author(s) | Title | Reference |
|---|---|---|
| Gordon, A. J. | 'A survey of atomic and molecular models' | J. Chem. Educ. **47**, 30–32 (1970) |
| Gordon, A. J. and Ford, R. A. | 'Atomic and molecular models' | Section in 'The Chemist's Companion', Wiley Interscience (1972) pp. 499–506. |
| Kakabadse, G. J. and Theobald, D. W. | 'Using models in chemistry' | Educ. in Chem. **4**, 135–138 (1967). |
| Nicholson, D. G. | 'Modeling molecules' | Chem. Eng. News **30**, 3164–7 (1952) |
| Nuffield Chemistry | 'Models and their uses' | Nuffield Chemistry: Handbook for Teachers, Longmans/Penguin (1967) pp. 202–233. |
| Savory, C. G. | 'A survey of crystal and molecular models' | Educ. in Chem. **13**, 136–141 (1976) |
| Smith, D. K. | 'Bibliography on molecular and crystal structure models' | Natn. Bur. Stand. Monogr. 14 (1960). |
| Walton, A. | 'The use of models in stereochemistry' | Prog. Stereochem. **4**, Ch. 8 (eds. B. J. Aylett and M. M. Harris), Butterworth (1969), pp. 335–375. |

### 1.2.4 Prices

In view of the current rate of inflation, it would be pointless to give the actual prices of models. However, some idea of the order of magnitude involved is conveyed by the price code given in Table 1.4.

**Table 1.4**
Price Code for Models

| Code | Description | Approx. price range |
|---|---|---|
| a | cheap | less than £10 |
| b | moderately cheap | £10–50 |
| c | moderately expensive | £50–100 |
| d | expensive | greater than £100 |

It should be remembered that in the U.K. Value Added Tax is charged on goods from EEC countries. Price codes are for the country of origin only, and duty is payable on imported goods. Freight charges can also be very considerable and some manufacturers impose a surcharge on small orders. Another problem is variation in exchange rates when importing goods. Wherever practicable, I have indicated in the lists at the ends of each Chapter who the distributors are in Western Europe and the USA, and readers wishing to purchase models are advised to get a firm quotation and a delivery date from a distributor in their own country whenever possible. The cost of imported models can be alarmingly high and apparently bear little relationship to the quoted catalogue price in the country of origin. Where there is a wide range of models (such as spheres) with prices varying from very cheap to expensive, a single price code would be misleading. In such cases more than one code is given. For example, code 'a,b' would mean that although prices go into the 'b' range, models at a low price are also available. Readers will appreciate that the problem of prices is a very complicated one at the moment. However, it seemed important to give some indication of cost because I know from experience that it is sometimes necessary to reject a model (or other equipment) on grounds of cost alone no matter how excellent its qualities. Availability of models fluctuates and agents for the distribution of foreign models change rather frequently. The information given is correct, to the best of my knowledge, at the time of going to press. If difficulties arise, readers should contact the manufacturers in the country of origin directly.

### 1.2.5 References and Bibliography

As is to be expected, a large number of articles on models are published in the two major chemical education journals, *Education in Chemistry* and the *Journal of Chemical Education.* On the assumption that these journals are reasonably accessible to most readers, no more than a brief indication of the contents of these articles is given. However it is hoped that this will be enough to enable the reader to judge whether he needs to consult the contents in full. The references are grouped into categories so that the reader should find information on a particular type of model in the same part of the text. Other considerations apart, the text would lose coherence and become inordinately long if all these articles were fully summarised. There are few references to papers in obscure journals, except where these might be of historical interest, because papers which are difficult to obtain are unlikely to be of much value to the majority of readers. My intention is either to present the information in the text or to indicate where it may *easily* be found. Additional bibliographies including more general works such as books and surveys are given at the ends of Chapters where it is thought that they might be of interest. Because it is difficult to make sharp divisions between categories, some references occur at the end of more than one Chapter. This has been done in order to avoid a complicated system of cross-references which might irritate the reader.

## 1.3 CHEMICAL CONSIDERATIONS: ATOMIC AND IONIC SIZE: CRYSTAL STRUCTURES

According to wave mechanics there is a finite probability of finding an electron anywhere between the atomic nucleus with which it is associated and infinity. Therefore the size of an atom cannot be defined *precisely* using electron density data. The boundary surface is usually fixed in terms of a 90 or 95% probability of the electrons being inside it. However, in practical terms the size of an atom depends on its environment and is determined by the closeness of approach of other atoms, ions or molecules. These may or may not form primary chemical bonds with it. Attractive forces exist between all chemical species but as they approach one another repulsive forces build up quickly until the attractive and repulsive forces exactly balance. The distance from the nucleus to the point where the opposing forces balance is therefore the radius of the chemical species in that specific situation. For nonbonded species the relevant radii are van der Waals (for solids) and collision (for gases), the latter being about 10% less than the former. Internuclear distances for nonbonded species are of the order 0.35-0.40 nm, [18] so in a scale model it is sometimes possible to decide whether primary bonding is likely by measuring distances between atom models in an assembly.

The interatomic distance between two covalently bonded atoms is equal to the sum of the covalent radii. Therefore the single homonuclear bonds in diatomic molecules half the interatomic distance (usually obtained from X-ray or spectroscopic data) gives a covalent radius, and this can then be used to obtain the radius of the other atom in a heteronuclear diatomic molecule. However bond multiplicity decreases the covalent radii. Although many radii for doubly and triply bonded species are known, problems may arise where the bond order or even the nature of the bonding is not accurately known, for example in *f* electron element carbides and similar compounds. If we imagine the nonbonded atom as a sphere having the van der Waal's radius, then when two such atoms form a covalent bond the strong forces of attraction enable them to approach nearer to each other than the van der Waal's radius before being balanced by the repulsion forces. It is possible to envisage flat faces being formed on the spheres as the electron cloud is compressed and these faces are in contact in the molecule. However, the full van der Waal's radius is retained at the 'other end' of each sphere. Hence the effect of the strongly directional character of the covalent bond is that compressions occur and plane faces are formed on each sphere perpendicular to the bonds formed. This process is reflected in the shapes of the atomic models described in Chapter 2.

Ambiguities occasionally occur in textbooks over such terms as atomic, van der Waals, covalent and 'metallic' radii. These are sometimes confused, and it is important to know to what the numbers in a table are referring [19]. While 'atomic' is presumably synonymous with 'van der Waals', as defined above, an atom is still an atom when it is part of a molecule and we have to consider its covalent radius in a particular bonding situation as well. The term 'atomic radius' has also been used interchangeably with 'covalent radius' for nonmetals, and the sizes of the constituents of metal crystals have been described in terms of covalent radii [20]. This can be confusing, particularly when an hexagonal close-packed metal crystal can have two different values for the radii of its atoms (or ions) depending on their positions. For example, the internuclear separation of two basal atoms in a zinc crystal is 0.133 nm while between two vertical layers it is 0.146 nm [19].

The stable *s* and *p* block ions mostly have filled shell configurations and, if the 'hard sphere' model were entirely applicable, it should be possible to define their sizes precisely. However, this is really only possible for the cations and smaller anions because the larger anions are easily polarisable and some distortion of the electron cloud results in the presence of a polarising cation. Therefore care should be taken in inferring size from simple geometry and radius ratio considerations. For a given element which is capable of forming stable cations in more than one oxidation state, the higher the charge, the smaller the cation. Another factor affecting ionic size is the co-ordination number. The same cation is larger when it has eight anion nearest neighbours than when it has six. Similarly the six co-ordinate cation is larger than the four co-ordinate one, the ratios being approximately 1.04:1:0:0.94 for [8]:[6]:[4]. Standard practice is to quote the size for six co-ordination in tables but this is not always made clear, particularly in older textbooks.

It is commonly stated that cations are smaller than the atoms from which they are formed while anions are larger. The first part of the statement is certainly true but if we take the van der Waals' radius as being the 'true' atomic radius then addition of one or two electrons to form an anion in fact makes very little difference to the size, a matter of one or two per cent only [21].

In addition to the above considerations, the sizes of the *d* block cations are affected partly by the overall contraction along the series as the inner *d* shell is filled. They are also affected by the extent of ligand or crystal field stabilisation energy, which depends on the electronic configuration of each ion. Where this is large, as it is for $d^3$ and $d^8$ ions in an octahedral field. the ions are smaller because the ligands are able to approach nearer to the central ion before the attractive forces are balanced by repulsions. Superimposed on this is the Jahn-Teller Effect. In a weak octahedral field, say, this will have the effect of reducing the sizes of $d^4$ and $d^9$ ions. There is a fairly regular reduction in size with increase in atomic number across both series of *f* electron elements. Again this is because an inner shell is being filled, which results in the outer electrons being more strongly attracted to the nucleus so that the ion contracts.

There are many sources of information on size and structural data. Perhaps the most comprehensive are the Landolt Börnstein Tables [22] and the Chemical Society Special Publications on interatomic distances [23]. Wyckoff's excellent series on crystal structures [24] gives useful summaries of essential data on each structure, while for minerals *The Structures of Minerals* by Bragg and Claringbull is strongly recommended [25]. Many structural data quoted by models manufacturers come from Pauling's classic treatise *The Nature of the*

*Chemical Bond* [26] or from Well's book *Structural Inorganic Chemistry*, now in its fourth edition [27]. Writers of books on models generally seem to give only one bibliography at, or near, the end of the book so that it is not always easy to pin down the source of a specific piece of information. Manufacturers' booklets, supplied with the models, usually either have a general bibliography or no references at all. It is hoped that the arrangement in this book will enable readers readily to identify the appropriate reference.

## 1.4 WHY A BOOK ON MODELS?

First, models are concerned with structure and one of the pleasures to be derived from a study of structural chemistry is the recognition of correlations between structure and function. This is chemistry at its most reasonable, and such relationships should be emphasised to new students of the subject until they are sufficiently advanced to be able to cope with apparently intractable chemical problems. Models provide a most satisfying means of demonstrating structure/function correlations. A great deal of important research in chemistry and related subjects is currently concerned with structures, particularly those of macromolecules, and here models are an invaluable aid to understanding (see, for example, Watson's *The Double Helix*) [28]. Moreover, a study of these molecules immediately fires the imagination of young people, and what better way is there of putting across structural features than a model when their knowledge of structural chemistry is slight? The structures of a range of simpler molecules and crystals are now studied in the schools and here again the subject can quickly come alive with models as aids. Models are *fun* and teaching is most effective when both teachers and students are enjoying themselves. Looking into the future, apart from uses in teaching and bio-organic chemistry it seems probable that models could be used more extensively in the study of surfaces. Although little has been published so far along these lines, this is an expanding research field with important practical significance and models of surfaces can certainly sometimes reveal unsuspected stereochemical features which can help to elucidate chemical processes.

A second reason for a book on models at this time is that there has been a proliferation of new types in the last decade or so. At a purely practical level some kind of consumer guide seemed desirable, especially as some of the newer models are particularly versatile and can be used in a variety of teaching and research contexts. Some of them are also cheap. When I first wrote about models, ten years ago [29], there seemed to be a need for cheap, fairly general purpose but adequate space-filling and skeletal models. These are now available, so clearly manufacturers think that models are a good commercial enterprise. The range of manufactured and homemade models is now very considerable and, to my knowledge, has not been surveyed in its entirety before. It is my earnest wish that teachers and research workers will be encouraged to experiment with unfamiliar model types and that they will find them useful.

Thirdly, my reason for writing this book is sheer self-indulgence. I like using models, talking about them and writing about them. They help us to understand chemical problems and to explain chemistry to others. They are satisfying to build and frequently the result is aesthetically pleasing.

Students like models. An unfamiliar model handed to a class for comment will be treated as a puzzle and 'solved' almost like an intellectual game. Models have a powerful relaxing and de-formalising effect in classes, particularly when they can be handled by the students and passed round for examination. The mere fact of handling a material object and passing it from one to the other helps to lower the barrier between teacher and student. Models are also valuable in 'trouble-shooting' sessions as they provide an excellent means of testing comprehension. Whenever possible they should be available for inspection outside the classroom or lecture room and, if there is space, it is often helpful to set them out on a bench during laboratory classes. There are often short periods during processes like evaporating, digesting, drying, distilling, cooling and so on when the student has time to examine the models. I believe that the duty of a teacher is two-fold. The first and major purpose is to educate, the second to present the material in such a way that the student will be able to assimilate it and ultimately reproduce it in examinations. Primarily models serve the first purpose, they educate. A two-dimensional drawing of a three-dimensional structure on the blackboard shows the student how to present the structure in an examination paper. Both are necessary.

Finally, I should like to borrow a couplet from W. S. Gilbert [30], much quoted by Dr. B. D. Shaw in his famous 'Explosives' lecture:

'For he who'd make his fellow-creatures wise
Should always gild the philosophic pill!'

## REFERENCES

[1] Pepys, S., *Diary,* entry on 30 July 1662. Edition edited by R. C. Latham and W. Matthews, Bell (1970).
[2] Kakabadse, G. J. and Theobald, D. W., *Educ. in Chem.*, **4**, 135 (1967).
[3] Theobald, D. W., *Educ. in Chem.,* **5**, 99 (1968).
[4] Clarke, D. L., *Models in Archaelogy,* Methuen (1973).
[5] Theobald, D. W., *Philosophy,* **39**, 260 (1964).
[6] Levine, F. S., *Educ. in Chem.,* **11**, 84 (1974).
[7] Langmuir, I., *J. Am. Chem. Soc.,* **51**, 2847 (1929).
[8] Petersen, Q. R., *J. Chem. Educ.,* **47**, 24 (1970).
[9] Hutchinson, E., *J. Chem. Educ.,* **45**, 600 (1968).
[10] Lippincott, W. L., *J. Chem. Educ.,* **45**, 1 (1968).
[11] Theobald, D. W. *An Introduction to the Philosophy of Science,* Methuen (1968).
[12] Davies, D. S., *Chemy Ind.,* 493 (1973).

[13] Ollis, W. D., *Proc. Roy. Inst.*, **45**, 1 (1972).
[14] Mislow, K., *Introduction to Stereochemistry*, Benjamin (1965).
[15] Herron, J. D., *J. Chem. Educ.*, **52**, 146 (1975).
[16] Beistel, D. W., *J. Chem. Educ.*, **52**, 151 (1975).
[17] Institute of Physics, *J. Scient. Instrum.*, **24**, 249 (1947).
[18] Baker, R. W. and Pauling, P., *Chem. Commun.*, 573 (1970).
[19] Nettleship, J. L. *Educ. in Chem.*, **2**, 241 (1965).
[20] Heslop, R. B. and Robinson, P. L., *Inorganic Chemistry*, Elsevier (1960).
[21] Nettleship, J. L., *Educ. in Chem.*, **5**, 156 (1968).
[22] Landolt, H. H. and Börnstein, R., *Tables*, New Series, group III, particularly vols. 5, 6 and 7, Springer (1950 *et seq.*).
[23] Chemical Society, *Spec. Publs.* nos. 11 (1958) and 18 (1965), ed. L. E. Sutton.
[24] Wyckoff, R. W. *Crystal Structures*, Wiley-Interscience (1960 *et seq.*).
[25] Bragg, L. and Claringbull, G. F., *Crystal Structures of Minerals*, Bell (1965).
[26] Pauling, L., *The Nature of the Chemical Bond*, 3rd edition, Cornell and Oxford (1960).
[27] Wells, A. F., *Structural Inorganic Chemistry*, 4th edition, Oxford (1975).
[28] Watson, J. D., *The Double Helix*, Weidenfeld and Nicolson (1968).
[29] Walton, A., *Prog. Stereochem.*, **4**, Chapter 8, edited B. J. Aylett and M. M. Harris, Butterworth (1969).
[30] Gilbert, W. S., *The Savoy Operas*, Macmillan (1963).

# 2

# Space-filling molecular models

## 2.1 INTRODUCTION

The idea of a molecule is itself a model but if we regard this concept as a reasonable approximation to reality then it is possible to visualise a molecule in our imaginations. In the gas-phase we picture each molecule rushing around, frequently colliding with others and exhibiting rotational and vibrational movements within itself as well. Even in a molecular crystal, vibrations about a mean position still occur. Therefore any attempt to portray a molecule other than in constant motion is already inadequate. However, if we disregard this molecular mobility and use a static model, then the space-filling variety is more like a scaled-up molecule than any other kind. In the best space-filling models the relative sizes, shapes, interatomic distances and bond angles correspond with literature values to a very good approximation and conform, as far as possible, to known chemical facts.

General principles of construction for most manufactured space-filling models are similar. Specific details are given later in this Chapter when particular types are considered. Initially, for each atom model a sphere is selected (according to the scale chosen) whose radius is proportional to the van der Waals' radius of the isolated atom. This sphere then defines the volume occupied by the atom and indicates the limit of tolerance to the approach of any other atom or molecule with which it does not form a primary bond. When a primary bond is formed, the attraction between the two atoms results in a reduction of the distance between them. The interatomic distance is then the sum of the covalent radii. In order to simulate such an effect in molecular models, segments have to be removed from the original spheres. These must be perpendicular to the bonding direction, such that the distance between the centre of the sphere and the centre of the plane face formed by removal of a segment corresponds to the covalent radius of that atom in the appropriate bonding situation. A hydrogen atom model would be derived from a sphere whose radius was proportional to the van der Waals' radius of hydrogen and one segment would be removed so that the distance from the centre of the sphere to the plane face would represent the covalent radius of hydrogen. An oxygen atom model which might be bonded with hydrogen to form a model of the water molecule would have two segments removed, inclined at the tetrahedral angle. Similarly an aliphatic carbon atom model would have four segments removed so that the bonding directions would be tetrahedrally disposed. This system is used for simple single bonds.

Models representing multiply bonded atoms have to be modified in several different ways. The distance between the linked atom models must be less than in the singly bonded species and the shape of the model and the symmetry of the linking must be changed. Moreover some mechanism must be found whereby rotation about the link is restricted or precluded, so simulating the effect of multiple bonding. For example, triply bonding carbon is usually represented by a flattish cylindrical atom model with a linking position at the centre of each of the flat faces, giving linear symmetry, and some means is provided for preventing rotation about the linkages. A further complication arises in representations of cyclic molecules. A benzene carbon model which is precise has approximately the shape of a trigonal prism. It must have two bonding faces which when linked with other similar atom models will have a distance between them appropriate to the $\pi$-bonding system in the benzene ring. A third face links with a hydrogen atom model to give, proportionately, the single bond C-H distance.

Another problem is that of linking rings in models of polycyclic systems, especially where the rings are of different sizes. Many manufacturers produce special models for this purpose. A consideration of the variation in bond angle and bond length in four-, five-, six-membered and larger rings, saturated or unsaturated, heterocyclic or otherwise, will give an indication of the large number of atom models, particularly of carbon, which are required for accurate representations of such systems. The producer of these models has two choices. Either he can produce a specific model for each reasonably probable application, at high cost, or he can produce a few compromise models which are never exactly suitable but are much less expensive. Approximately equal numbers of the models discussed in this Chapter fall in each category.

By far the most important *specific* use of space-filling molecular models lies in their ability to indicate the existence of steric hindrance. In order to avoid spurious indications of steric hindrance in molecular models, some atom models have had to be further modified so that the whole assembly is in accordance with known chemical facts. In some cases arbitrary reductions are made in radii, while in others the larger atom models are either chamfered around the edge of the plane face or truncated if sharp corners occur.

Apart from the atom models, the other major component in space-filling models is in the linkage mechanism. Considerable variation and subtlety is possible in the mode of linkage of atom models. Ideally it should be possible either to shorten or lengthen the link, to effect small changes in bond angle, and to control the extent to which rotation about the linkage is permitted. The designers of the manufactured models have produced a number of mechanisms by means of which these changes may be brought about. Each is described under the appropriate heading when particular models are discussed. Another factor which must be taken into account when the utility of a particular model type is being assessed is the ease of assembly and disassembly. Ease of manipulation is determined largely by the type of connector used, by whether tools are provided and, if so, by how efficient they are. Again, a wide range in facility of operation exists.

The use of space-filling models both in teaching and research are manifold [1], though in the former case there is considerable overlap with information which can be imparted by less complex (and cheaper) models. However, they demonstrate better than any other model our theoretical model of a molecule, and in more advanced teaching they may be used to demonstrate steric hindrance. The last phenomenon is the one for which space-filling molecular models are chiefly used in research but many other uses have been reported. Among the most important of these are the prediction of transition state intermediates and so the most likely reaction path and inference of probable conformations and correlations with coupling constants and other n.m.r. data. Calculation of molecular volumes and the cross-sectional area of sorbed molecules from molecular models is also important, as is estimation of probabilities of formation of inclusion compounds from the dimensions of the potential guest molecule as indicated by a space-filling model. Although this must not be taken too far, there is frequently an analogy between mechanical strain in a model and electrostatic repulsions in a molecule. Consequently when a molecular model is strained or is difficult or impossible to assemble it is often reasonable to infer that an alternative structure is more feasible. The use of space-filling models in the study of co-ordination complexes has also been described. The assembly of a suitable model may sometimes indicate the possibility of co-ordination between a donor atom and a metal and suggest the type of co-ordination. For example, it may indicate whether $\pi$-bonding is possible from the orientation of donor and acceptor. In this context, the possibility of co-ordination between solute and solvent in solution chemistry has also been clarified by the use of space-filling molecular models.

The models described are manufactured, with one exception, either in Western Europe or the U.S.A. Some are marketed as sets or kits of models which contain a number of different components for the assembly of a reasonably wide range of molecular models. Others are marketed as components.

## 2.2 STUART-BRIEGLEB SPACE-FILLING MODELS

Models of this type, designed by Stuart and modified by Briegleb, have been well-known and widely used in chemical circles for several decades. The atom models are constructed according to the principles outlined above and joined by connectors which bring the plane faces of the atom models virtually into contact. As a result the completed molecular model tends to be rather inflexible. Usually little angular distortion or bond length variation is possible. Many manufactured sets of models are of this type, while most descriptions of the preparation and assembly of homemade space-filling models give Stuart-type examples which may be designated Stuart, Briegleb or Stuart-Briegleb [2].

**Table 2.1** – Stuart-type space-filling molecular models

| Name | Price Code (Table 1.4) | no. of species / no. of elements | scale | material colour, density | chief applications | precision ‡ |
|---|---|---|---|---|---|---|
| Catalin | b,c | 53/16 | 10 mm≡0.1 nm | solid self-coloured phenolic resin | organic, organometallic some inorganic | ** |
| FHT | b | 29/12 | 10 mm≡0.1 nm | solid, self-coloured | organic, organometallic | ** |
| Godfrey | b | 26/16 | 16.5 mm≡0.1 nm | hollow, self-coloured flexible PVC | organic | * |
| Griffin and George | b | 12/6 | 15 mm≡0.1 nm (covalent) 12.5 mm≡0.1 nm (van der Waals) | hollow, self-coloured light plastic, rigid | simple organic or organo-metallic (no aromatic) | * |
| Leybold | b | 47/17 | 15 mm≡0.1 nm | solid wood, painted or hollow self-coloured plastic | biochemical, organometallic, organic, simple inorganic | *** |
| Molecular Fragments | a,b | 25/12 | 16.5 mm≡0.1 nm | high density, expanded polystyrene: may require painting | organic, organometallic biochemical some inorganic | * |
| S.A.S.M. | b,c,d | 52/16 | 10mm, 15 ,, 20 ,, 25 ,, ≡0.1 nm | painted wood, plastics or metal components | organic, organometallic relatively good for inorganic | ** |
| S.R.M. Demon-stration | c | 14/7 | 33 mm≡0.1 nm | high density, expanded polystyrene coloured | organic | * |

‡3, 2 or 1 stars equivalent to very good, good or fair.

## Some Manufactured Stuart Models

### 2.2.1 Catalin Covalent (Figure 2.1).

These models are usually sold as sets but standard packs of a particular atom model may also be purchased. A total of fifty-three different atom models representing sixteen elements are manufactured. Thirteen atom models repre-

| strength of assembly/ease of manipulation ‡ | connectors | rotation/ nonrotation | linkage length variation | linkage angle variation | H-bonding mechanism | special models or groups |
|---|---|---|---|---|---|---|
| **/*** | tubular rubber pegs | free or prevented as appropriate by pin and socket mechanism | only by partial insertion of link | approx 5° | none, donor atoms joined by wire | 5 metal, silicate oxygen |
| ***/*** | snap-fastener (press-stud) | free | none | slight | none | one metal, co-ordination species (2 nitrogen, 2 oxygen) |
| */* | flat, barbed, polythene strips | free if rod used, prevented by stapling where necessary | only by partial insertion of link | large, very flexible assembly | drawing pin in one model attracted to magnet in another | |
| **/*** | plastic tubes of round or square cross section | free with round links, prevented with square | only by partial insertion of link | slight | none | 6 co-ordinate metal |
| ***/** | snap-fastener (press-stud) | free or prevented by clamping springs where appropriate | none | slight | 'hook-and-eye' | wide range for cyclic systems, base-pairs small rings, groups |
| */* | wooden pegs which can be glued or inserted into vinyl collars | none if glued, free if in vinyl collar | only by partial insertion of link | slight | special H-bond hydrogen | octahedral metal atom |
| **/*** | various lengths → 'exploded' models; metal or plastic | single link for free, double links for restricted | only by partial insertion of link | slight | linear, nonlinear and intermolecular | 9 metal, various symmetries 2 H-bond |
| ***/*** | snap/fastener (press-stud) | free | none | slight | none | benzene C=C electron pair |

sent carbon while there are eight each of oxygen and nitrogen representing these elements in varying environments in an organic molecule. Therefore this set, like many others, is particularly well suited to the construction of models of organic compounds and is quite versatile for this purpose. However, because five metal atom models are available and an oxygen atom model with its links disposed at a 140° angle, the Catalin Covalent set may be used to assemble models of organometallic compounds and silicates as well as some simple inorganic compounds. These can also be used in conjunction with the Catalin Ionic set (Chapter 3) which is to the same scale (10 mm ≡ 0.1 nm). For example, I have studied the stereochemical relationships between a crystal surface and adsorbed molecules in this way.

Fig. 2.1 – Catalin covalent: (a) carbon tetrachloride, (b) *n*-pentane, (c) tyrosine, (d) neopentane.

The atom models are made of self-coloured, moulded, rather dense plastic. The manufacturers have selected the size so that, proportionately, the van der Waals' radius is reduced by 20%. This, they claim, precludes spurious steric hindrance effects and allows some indication of interpenetration of electron clouds. It is this reduction which permits the construction of three and four-membered rings. However an overall reduction of this type might fail to indicate steric hindrance in the model where it occurs in the molecule. Block rings are available for benzene and naphthalene which save time in connecting the constituent atom models. These have holes at the centre into which rubber mouldings or some other material may be inserted to indicate the extent to which the $\pi$-cloud extends above and below aromatic rings. There is a 'distorted benzene' carbon model which represents the atom common to linked six- and five-membered rings. A special carbon atom model is also used in four- or five-membered rings.

The connectors are made of short lengths of fairly flexible rubber tubing which are inserted into the atom models by means of the pliers provided. These can also be used to extract them. The rubber pegs permit about 5° angular distortion from the linear when two atom models are linked and are therefore almost touching each other. The molecular models are very easy to assemble and disassemble but, as a result of the relatively high density, a large model requires extra mechanical support and the manufacturers recommend passing wire through the whole assembly. This is possible with tubular connectors and a little extra drilling inside the atom models. Although rotation about the links can occur when the bond represented is single, when it is multiple rotation is prevented by inserting a steel peg on one atom model into a socket on its neighbour, in addition to the usual tubular connector.

A defect of the Catalin Covalent system is the absence of a specific model for hydrogen bonding. Unfortunately recourse to wire is again necessary in order to join the two donor atoms.

Longer (100 mm) steel connectors may be used in the assembly of 'exploded' models. These have the advantage of retaining the space-filling characteristics of the functional group but, for example by lengthening the connector linking the group to an aromatic ring, greater clarity may be achieved.

The Catalin Covalent models are quite precise and easy to manipulate. with a small scale and a good range of atom models. They are recommended as being the nearest to a 'general purpose' set which is still reasonably precise among the Stuart-type models.

### 2.2.2 Fisher-Hirshfelder-Taylor (FHT)

These are in many respects, including the scale, similar to the Catalin models. However they have snap-fastener links (like press-studs) making rotation about the single link very free. A connector lifter is used to separate the strongly linked atom models.

The Organic Atom Model Kit contains eighty-six models of twenty-nine types representing twelve elements. The Metal Co-ordination Kit, for the construction of models of metal complexes, contains eighty-seven models of fourteen different types representing seven elements. The range of atom models offered is not as wide as that provided by Catalin and there is, again, no hydrogen bond model.

The colours used depart from the usual convention in that hydrogen atom models are orange, oxygen light blue, bromine tan and iodine brown.

### 2.2.3 Godfrey Molecular Models (Figure 2.2).

The Godfrey Model Kit contains one hundred atom models with eleven different types representing six elements, but additional atom models can be obtained bringing the total to twenty-six different types (sixteen elements in all). The atom models are hollow and of self-coloured, flexible PVC. It is claimed that the flexibility and compressibility removes the need for a large number of each element [3]. Therefore there are only three carbon types. Tetrahedral (aliphatic) carbon atom is represented by a sphere with a radius proportional to the covalent radius (tetrahedral nitrogen is also represented by a sphere of this type). Trigonal carbon, used in any ring system or for double bond carbon, is represented by a capsule-shaped model, and for linear carbon the model is disc-shaped and represents triple bond carbon. The flexibility enables strained structures, such as small rings, to be built with relative ease.

Fig. 2.2 – Godfrey space-filling: (a) tyrosine, (b) carbon tetrachloride, (c) hydrogen bond.

The connectors are flat strips of polythene in four different lengths, approximately 12.5, 26.5, 40.5 and 51.0 mm long. The shortest is barbed at both ends and used where the atom model is connected to more than one other, except in the construction of ring systems where the connectors are half inch lengths of plastic rod. The three longer ones are used with models representing monovalent species because these each only have one 'bond receptacle' This is a hole, thickened round the edges to minimise wear and tear, into which the connector can be inserted. The nonbarbed end is pushed through the 'bond receptacle' until it touches the inside surface of the atom model. The projecting barbed end is then attached to an adjacent atom model.

The set is designed for the assembly of models of organic molecules but a few simple inorganic ones can be constructed. The scale is approximately 16.5 mm ≡ 0.1 nm but only moderate accuracy can be achieved because the models tend to distort with use. The original shape cannot always be completely restored by immersion in boiling water, as is claimed, but this procedure effects a considerable improvement. Assembly of these molecular models requires some manipulative skill because sometimes the connectors tend to become disengaged before the model is completed. Increased mechanical stability may be brought about by inserting a bridging staple into adjacent atom models. This is recommended for ring systems and also as a means of preventing rotation about a link where appropriate.

Hydrogen bonding is simulated by attaching one donor to a hydrogen atom model in the normal way and then pressing a drawing-pin into the rounded surface of the hydrogen. The second atom model to which the hydrogen is to be 'bonded' is then slit so that a small magnet may be inserted. The magnet attracts the drawing-pin and this mechanism is said to demonstrate the electrostatic nature of the hydrogen bond.

### 2.2.4 Griffin and George Stuart Models

The atom models are hollow and moulded from light weight, coloured plastics. Although the boxed set contains one hundred and sixty-eight atom models, there are only twelve different types representing six elements and there is no aromatic carbon species. Oxygen atom models are again light blue and hydrogen orange or white. Their chief use is the assembly of fairly simple organic or organometallic molecules (a six-co-ordinate metal atom is included) and they are designed solely for educational purposes. Only a moderate order of accuracy is claimed for them. The scales differ for the covalent and van der Waals' radii: for the former 15 mm $\equiv$ 0.1 nm while for the latter 12.5 mm $\equiv$ 0.1 nm approximately.

Links of circular section (one hundred) and square section (thirty) are provided in the set and some atom models have one or more links moulded in so as to save assembly time. The links are also of hollow, moulded plastics. The round ones are used for singly linked atom models and the square ones represent multiple bonds. Thus rotation about the single link is possible, although the frictional resistance is quite marked, and no rotation is possible where the square links are used.

These models are simple and easy to use, requiring no tools. They are suitable for elementary teaching and are cheaper than most other space-filling models.

### 2.2.5 Leybold Molecular Models (Figure 2.3)

The number of atom models and the type varies with which of the four kits manufactured is chosen but components can be bought separately. A good range is available consisting of forty-seven different models (including two base pairs for the construction of DNA models) and with seventeen elements represented. The Leybold models are particularly well-suited to the construction of biochemically important species including biologically active metal chelates. Many simpler models of organic molecules and some inorganic molecules can also be assembled. For the construction of models of the various types of polycyclic and heterocyclic ring system which occur in biochemical structures a number of highly specialised species must be available, and the Leybold models with twelve carbon and ten nitrogen atom models cover the range required up to and including the DNA molecule. Included are both twin atom models which link fused ring systems and a carbon bridge for use in six-membered chelate rings. Other models of interest are special carbon atoms for three- and four-membered rings and allene, and also groups such as epoxide and nitro. Preassembled base pair groups, cytosine-guanine and adenine-thymine, are also available.

Fig. 2.3 – Typical Stuart models: (a) rubber Czechoslovakian model of acetic acid, (b) Leybold tyrosine, (c) Leybold hydrogen bond, (d) Leybold trichloromethane.

Originally the models were made of wood and painted but now some are made of self-coloured plastic and are hollow. These are precise models, the scale being 15 mm ≡ 0.1 nm. With the exception of halogens and phosphorus the colours conform to the usual convention, though where particular faces need to be distinguished a second colour is sometimes used. For example, carbon atoms for use in five-membered rings are black with two yellow faces. The yellow faces must be linked to similar ones to form the ring.

The atom models are attached to each other by strong snap-fastener (press-stud) links. They are easily assembled simply by pushing them together but usually require the separating tool provided to pull them apart. This tool can be used to insert and remove the links. Free rotation occurs about the link so readily that some extra support may occasionally be necessary to retain an assembly in the required conformation. Where free rotation does not occur in the molecule, adjacent atom models have aligned slits into which clamping springs are inserted. These also confer extra stability. An excellent 'hook-and-eye' device is used to demonstrate hydrogen bonding. There is a special hydrogen atom model for this purpose which is flatter than the normal one and which has a hook attached to its surface. This hook can be inserted into a slit spanned by a metal staple on the donor atom. It is usually a fairly tight fit and the resulting assembly is very firm, a highly desirable feature in large models. The only disadvantage is that little variation in orientation is possible. This is generally true of the Leybold models because they are true Stuart models and adjacent model atoms when linked are almost in contact. The nature of the link precludes any variation in length and angular distortion can only be very slight.

However, given the limitations of the Stuart type these are good precision models and very satisfactory assemblies of varying size and complexity can be built from them.

Some other models with the same scale and very similar to the Leybold ones were designed and produced at the University of Chemical Technology, Prague, Czechoslovakia. These are of hard, appropriately coloured rubber with dumb-bell shaped links of solid metal which can be pressed into holes in the atom models. They are described by the designers as Briegleb models.

### 2.2.6 Molecular Fragments (Figure 2.4).

The scale of these models is the same as that for the Godfrey models (16.5 mm ≡ 0.1 nm) and some of the atom models are similar in shape. However, they are made of high density expanded polystyrene. The basic student set contains sixty units with atom models of twelve different shapes and sizes, some existing in more than one colour to give a total of twenty-five variants. It is possible to buy all the shapes in white and paint them oneself, either with the manufacturer's colour kit or with commercial paints. Apart from hydrogen atom models, the student set contains coloured versions of all the other types.

Several of the atom models are of particular interest, among them H-bonding hydrogen, which is flattish with linearly disposed links. A small cubo-octahedron with tetrahedrally arranged links in black or blue represents tetrahedral carbon or nitrogen, and a sphere larger tetrahedrally bonding atoms such as phosphorus, silicon or sulphur. An octahedral shape represents a metal atom, which could have tetrahedrally disposed links as well as octahedral symmetry. Apart from the hydrogen bonding model, all these are of a size corresponding *only* to the covalent radii. Other atom models conform to the usual Stuart design.

Each is marked at the point (or points) where linkages should be made. Using a pointed wooden dowel, for example, holes should be punched at these places so that the vinyl collars provided with the set can be inserted and glued in position. Of course, it is not essential for vinyl collars to be used if the assembly

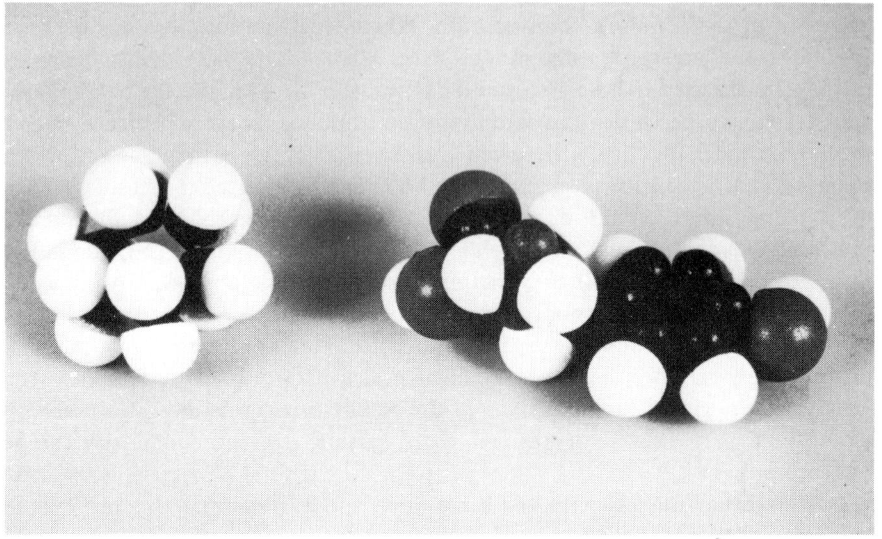

Fig. 2.4 – Molecular Fragments: (a) cyclohexane, (b) tyrosine.

is to be permanent and if free rotation about a link is not required. However, for models which are to be assembled and dismantled many times the manufacturers recommend the use of these collars and subsequent linkage of atom models by insertion of hardwood pegs into them. The peg can be used alone and glued in position if preferred, for example when rotation should not occur. A further advantage in using vinyl collars is that a little more flexibility in the assembly is achieved. 'Molecular Fragments' are more versatile than some models already described because variations in bond angle in a particular molecule can be demonstrated simply by changing the position of the link. Changes in radii can be simulated by sanding down the model to make it smaller where appropriate and a variety of paints, glues and connectors can be used.

Various other kits are available in addition to the basic student set and from these may be built, for example, protein, polypeptide or DNA models. Preassembled constituents are available such as peptide or amino acid units and appropriate side chains. A space-filling model of DNA can be obtained either preassembled or unassembled.

These models, though less sophisticated than many manufactured models, have the advantage of being made of light, tractable material which can be fabricated easily. They are versatile and inexpensive.

### 2.2.7 S.A.S.M. Models (Figure 2.5)

The Modèles Moléculaires Universels cover a wide range as regards number of different types of atom model produced (fifty-two with sixteen elements represented and, six different types of metal atom model in addition), scale, methods of assembly and presentation. Many different sets are available both for teaching and research purposes.

The atom models exhibit several interesting features. They are essentially of the Stuart type with minor variations. Only six different radii correspond to the covalent radii so that, for example, carbon, nitrogen and oxygen have the same, as also do chlorine, silicon, phosphorus, sulphur, arsenic, tin, titanium and chromium. However, there are about twenty variants on the van der Waals' radius. In addition to the metal atom models mentioned above, six others represent linear, tetrahedral, square planar, trigonal bipyramidal, square pyramidal and octahedral symmetries. Moreover, halogen atom models designed to represent these elements in inorganic species such as oxyanion or interhalogens make these space-filling models more useful to the inorganic chemist than many others, although it would be fair to say that the emphasis is still on organic and organometallic chemistry.

Fig. 2.5 – SASM: (a) glycine, (b) tyrosine.

The colour scheme used is not the one generally accepted in Britain and the USA (Table 1.1). Hydrogen atom models are red and oxygen blue, the halogens are yellow, green, brown and violet, and phosphorus is grey-brown.

A useful feature of these models is that hydrogen bonding is catered for by two hydrogen atom models, one for linear H-bonds and the other for intramolecular H-bonds. In addition there is a special oxygen atom model for intermolecular, nonlinear H-bonds such as are found in biochemical systems.

The atom models can be obtained in four sizes, the scales being 10, 15, 20 or 25 mm ≡ 0.1 nm. The connectors are also produced in lengths corresponding to scales varying fron 10 mm to 100 mm ≡ 0.1 nm. Only by using the shortest connectors do true space-filling models result, that is with the faces of the atom models virtually in contact as in all Stuart-types. Using longer connectors effectively converts the models into the ball-and-spoke type (see Chapter 4) but it can be advantageous to retain space-filling characteristics in a functional group substituent in an aromatic ring, say. Steric hindrance may indicate a possible reaction path in such a situation but by using an 'exploded' model of the aromatic ring it is claimed that greater clarity is achieved. In 'exploded' models the scale for the atom models differs from the overall scale owing to the greater length of the connectors (see Catalin Covalent).

The connectors are made of metal or flexible plastic, the latter being used for strained structures and for representing multiple bonds where more than one connector is used to join two atom models. The connectors are pressed into apertures in the atom models and where strain is expected to be considerable they are dumb-bell shaped so as to fit very tightly.

The largest size models (scale 25 mm ≡ 0.1 nm) are suitable for demonstration purposes in a large lecture hall.

### 2.2.8 Science Related Materials (S.R.M.) Demonstration Size Models (Figure 2.6)

The scale of these models is 33 mm ≡ 0.1 nm but in spite of their size they are very light because they are made of high density expanded polystyrene. There are thirteen different atomic models representing seven elements and coloured according to the Institute of Physics code except that fluorine is

Fig. 2.6 – SRM demonstration size: cyclohexane.

orange. The shapes are similar to the Godfrey atom models. Although in older sets tetrahedral carbon and nitrogen were spherical with radii proportional to the covalent radii, in the current one these models are polyhedra with links in four tetrahedrally disposed plane faces. The set also includes a complete benzene ring model and another representing a carbon-carbon double bond. A model which can be used to indicate the position of an electron pair was included in the older set but is not in the newer one.

Snap-fastener connectors fixed to hardwood inserts join the atom models together forming a strong link about which free rotation can occur. The models are easy to manipulate and the completed assembly is very stable.

The models are limited in scope but make valuable teaching aids for large classes up to and including first year university level.

## 2.3 NEW DESIGN CONCEPTS IN SPACE-FILLING MOLECULAR MODELS

Within the last decade or so some new manufactured space-filling molecular models have been marketed incorporating features which represent significant departures from the Stuart-Briegleb concept. Two of the four types discussed retain in the atom model the full value of the radius proportional to the van der Waals' radius, but have that distance representing the covalent radius reduced. The balance is incorporated into the connector. Consequently in one type of model the equivalent of 0.05 nm of the interatomic distance resides in the link itself. Several advantages accrue, greater clarity is achieved in large assemblies, for example. Again greater angular distortion is possible, the length can be changed either by using an adjustable link or having several different lengths to substitute and the link can be used to effect free or restricted rotation where required.

Though much less sophisticated than the others, the third type is an entirely new design for space-filling models. Instead of a segment of a sphere being removed, part of it is covered by a hollow bowl so that when a series of bowls are joined the resulting molecular model looks similar to the traditional ones.

The models are described in order of appearance in their present form on the British market.

**Table 2.2**

Some new designs in space-filling molecular models

| | Courtauld | CPK | Scale Atom |
|---|---|---|---|
| Price Code (Table 1.4) | b | c,d | a,b |
| no. of species / no. of elements | 35/13 | 31/12 | 18/8 |

**Table 2.2** (*Cntd.*)

| | Courtauld | CPK | Scale Atom |
|---|---|---|---|
| scale | 20 mm≡0.1 nm | 12.5 mm≡0.1 nm | approx. 10 mm≡0.1 nm |
| materials, colour density | hollow, self-coloured plastic; light, rigid | hollow, self-coloured plastic; light rigid (except H-flexible) | hollow, self-coloured plastic; light, rigid |
| chief applications | biochemical, organic organometallic, some inorganic (eg silicates) | biochemical, organic organometallic, some simple inorganic | organic, biochemical, simple inorganic |
| precision‡ | *** | *** | * |
| strength of assembly‡ / ease of manipulation | ***/** | ***/** | **/*** |
| connectors | brass cylinder press-stud + neoprene collar and (usually) a plastic ring | tough, resilient elastomer, 6 types, e.g. short, long, locking | mostly round pegs, square for multiple bond C or N, integral part of atom model |
| rotation/ nonrotation | free for single links; double links or nylon rods inserted into opposing atom models prevent it where appropriate | free or prevented by locking link where appropriate | usually free, prevented by square peg or locking pin |
| linkage length variation | adjustable links with screw thread give either up to 0.01 nm increase or up to 0.01 nm decrease | short or long links reduce or increase length by equivalent of 0.005 nm or 0.008 nm respectively. | none |
| linkage angle variation | approx 12°, 6° or 0° depending on collar and/or ring used in link | approx. 8° | slight |
| H-bonding mechanism | special hydrogen linked by elastic bands to donor | 2 H-bond; permanent (linear), bayonet type (linear or non-linear) fits into slots on donor atom | |
| special models or groups | 2 metal, 'silicate' oxygen, 'co-ordination' oxygen, H-bond, allene small rings, block rings, 5-ring nitrogen | several for fused rings 3 metal (one 'all-purpose'), H-bonds, special 4-ring carbons and nitrogens | C and N for fused rings, C and N for unsaturated 5-rings |

‡3, 2 or 1 star equivalent to very good, good, or fair.

### 2.3.1 Courtauld Models (Figure 2.7)

Dr. G. S. Hartley of Courtaulds Ltd. published the principles employed in the construction of these models in 1952 [4]. From then until 1966 several major developments took place, but in recent years the changes have been only minor [5].

Fig. 2.7 – Courtauld: (a) *trans*-dichlorobis(ethylenediamine) cobalt(III) cation (*trans*$[Co(en)_2Cl_2]^+$), (b) aniline, (c) nonlinear hydrogen bond, (d) linear hydrogen bond.

The thirty-five atom models representing thirteen elements were designed chiefly with the assembly of models of biochemical species in mind. However the inclusion of a metal atom model with octahedrally disposed links and of a copper one with tetrahedral symmetry enables some models of simple metal complexes and various organometallic compounds to be assembled. There is also an oxygen atom model with two linking sockets inclined at 141°, and silicate models can be built using this. Aquo co-ordination complexes are provided for with another such model. A special model is supplied for a hydrogen bond between, for example, two oxygens. This is joined to one by the usual link (described below) and attached to the rounded surface of the other (a carbonyl oxygen model) by three elastic bands. Each of these bands links a peg on the flat face of the H-bond model to a similar peg on that of the carbonyl oxygen. For depicting linear H-bonds, the face of the hydrogen model in contact with the round surface of the oxygen bears a vulcanised fibre strip which is removed when it is required to demonstrate a nonlinear bond. There is an allene carbon model and block rings for benzene and naphthalene. In most cases the equivalent of the full van der Waals' radius has been used but in the hydrogen atom model it has been reduced from the equivalent of 0.12 nm to 0.10 nm. It is claimed that this corresponds more closely to known chemical facts and that with a larger model spurious steric hindrance is observed. For a similar reason the halogen atom models (except fluorine) have been chamfered around the edge of the plane face so that hexahalo aromatic ring models can be constructed. As a result, however, no bond angle distortion is observed in models of *ortho*-halogen substituted aromatic rings although this occurs in the molecules. A special nitrogen atom model is available for five-membered heterocyclic rings so that models of the purine bases, and hence DNA, can be constructed. The graphite carbon model has to be used for the common atom between two fused rings but this is a reasonable compromise.

The models are hollow and of light, self-coloured plastic. There is an embossed code mark on each, and the faces are distinguished by these where there may be ambiguity. For example, the two faces of a benzene carbon model which are incorporated into the ring must not be confused with that forming a $\sigma$ bond with hydrogen.

The normal link consists of a brass cylinder with knobs at each end which fit tightly into press-stud sockets on the atom models. Its length is 10 mm, the equivalent of 0.05 nm because the scale is 20 mm $\equiv$ 0.1 nm. A magazine is provided for insertion and extraction of the links as it is sometimes difficult to do these operations with the fingers. The link is never used on its own. In a normal linkage between two atom models, the metal link is first covered with a neoprene collar and then a yellow plastic ring before completing the connection. Such a device increases the tensile strength of the assembly and allows an angular distortion of about 6°, which is chemically feasible.

If a red ring is used instead of a yellow one, hardly any angular distortion is possible. However, there are some situations where greater distortion (up to 12°) is required, for example in the construction of a five-membered ring. In these cases the neoprene collar is used alone. The resilience of the collar causes the atom models to assume positions perpendicular to the link when the strain is removed. If it is necessary to fix atom models at an angle to each other, one of the red rings can be cut in half in a plane which is not quite that of the ring. The two uneven rings so produced can be placed over the neoprene collar and rotated with respect to each other so that the double ring is in part wider and in part narrower than the original. In this way the atom models can be held at an angle which is a few degrees less than 180°. With some specially prepared bent links for three- and four-membered rings only the neoprene collars are used. There is evidence that bent bonds occur in small rings so such a device seems to be theoretically justified [6].

Small changes in the distance between models (and, by analogy, the bond length) can be brought about by use of an adjustable link with a screw thread. There are two of these, one reducing the length by the equivalent of up to 0.01 nm, the other increasing it by up to the same amount.

The models can be assembled and dismantled manually (apart from the insertion and extraction of links) and the normal single linkage permits free rotation although there is enough frictional resistance to ensure that a required conformation is retained. Double links are used for multiple bonds and although the manufacturers advise use of a collar and yellow ring for each link, I have found that it is easier to press links into two sockets simultaneously if the yellow ring is removed from one. There are instances where rotation is known to be restricted about a *single* bond. Where this occurs, rotation about the link between atom models can be prevented by insertion of a nylon rod into sockets on opposing faces of the atom models. For example, the benzene model is built with alternating double and single links and rotation in the latter case is prevented by insertion of a nylon rod parallel to the link. This device can also be used to lock an assembly in a particular conformation. For example, linked tetrahedral carbon atom models can be locked in the staggered conformation.

Courtauld models are precision models in which the type of linkage permits a much clearer picture of bonding in the molecule to be envisaged. Steric hindrance is indicated particularly well because the surface of the model is an accurate portrayal of the van der Waals' envelope.

### 2.3.2 CPK (Corey-Pauling-Koltun) Models (Figure 2.8)

These models were designed by a sub-committee of the U.S. National Institutes of Health, and their development was supported financially by the National Science Foundation and implemented by the American Society of Biological Chemists. They are based on the older Corey-Pauling models [7] with new connectors designed by Koltun [8].

The major principles of design are similar to those of the Courtauld models in that the CPK models are a modified version of the Stuart-type, with some of the distance representing the sum of the covalent radii in the link. Consequently the atom models are not in contact, giving overall 'transparency' and a clear picture of the van der Waals' envelope, the full radii being used. There are twenty-seven atomic species representing twelve elements. They are hollow and moulded from self-coloured light, rigid plastic, with the exception of the hydrogen model which is of compressible polyethylene. Each atom model is labelled and an identifying symbol is stamped on the face in cases where ambiguity may occur, for example with the orientation of a benzene carbon model. These are also precision models and the manufacturers claim that angles are correct to within $0°30'$, covalent radii to the equivalent of 0.001 nm and van der Waals' radii similarly to 0.003 nm. The scale is 12.5 mm $\equiv$ 0.1 nm.

Although models of metal complexes and many simple inorganic and organic compounds can be constructed, these models were designed primarily for building biochemical structures. For example, there are special twin carbon models for fused 6:5 and 6:6 rings, nitrogen models for both five- and six-membered rings and two H-bond models which can be attached to the surface of specially designed oxygen or nitrogen atom models.

Fig. 2.8 – CPK: (a) carbon tetrachloride, (b) tyrosine, (c) hydrogen bond.

For strong, permanent H-bond links a hydrogen atom model bearing a hook is linked by its normal permanent link to, for example, one oxygen model. The hook is inserted into one of three parallel slots in the rounded, 'nonbounding' surface of another oxygen atom model and then tightened so that it cannot slip out. One or two small discs called 'H-bond spacers' may be inserted between hook and slot to ensure that the correct distance is maintained. If the link is to be made to a nitrogen atom model then the 'amine cap', which converts a tetrahedral nitrogen to an amine model, has to have a screw passed through the hole at its centre. This is only possible after removal of its link. The hook is removed from the H-bond model also leaving a hole into which the screw is screwed. For large assemblies the hook H-bond model may be necessary but for most purposes the 'bayonet' type H-bond is much more satisfactory and easy to use. This has a normal permanent connector on one side while the other side bears a peg with a projection at the end. This can be pushed into one of the slots on an adjacent oxygen atom model and it is then held firm under slight tension. A further advantage with this model is that a nonlinear H-bond system is possible if the bayonet connector is inserted into one of the side slots.

The halogen atom models (except for fluorine) are chamfered like the Courtauld ones. There are three metal atom models, two having octahedrally disposed sockets and the same radius (equivalent to 0.132 nm) but being described as covalent and ionic models. They are different in shape, the ionic one being spherical. The third, called the 'All Purpose Metal Atom', is also spherical. It has a diameter corresponding to 0.27 nm with lines of latitude and longitude embossed at 30° intervals and an indentation at one pole. These markings enable the special connectors provided with this model to be arranged in virtually any required orientation.

Recently some new CPK atom models have become available. These are specifically designed for small rings, for example four-membered rings containing carbon or nitrogen such as occur in penicillin and cephalosporin antibiotics [9]. The new species comprise two carbon atom models, one of which is tetrahedral and the other trigonal, and two analogous nitrogen atom models. With these the range of structures which can be assembled from CPK models is significantly extended.

As with the Courtauld models, special interest attaches to the nature of the link. The CPK links are made of hard rubber-like elastomer which is resilient and flexible. Each has three projecting bands round it. The two at either end can be pushed into sockets on atom models leaving the middle band to act as a spacer between them. This is a strong linking system and large assemblies are stable. Rotation can occur where it is appropriate but there is sufficient friction even for pendant side chains to be maintained in the correct orientation. A tool is required for insertion and extraction of the links. This has a well at one end into which the link is placed for insertion, by twisting, into an atom model. The other end of the tool is a fork-ended levering blade used to prise the models apart and then to lever out the link. The models can be pressed together manually once one end of the link is inserted. Maximum angular distortion is about 8°. There are six different colour-coded links. The standard is off-white, while the short (blue) reduces the distance by the equivalent of 0.005 nm and the long (red) increases it by the equivalent of 0.008 nm. The locking (grey) has projections which fit into slots in the sockets preventing rotation where appropriate, and the carbon (black) is used for demonstrating restricted rotation about the carbon-carbon single bond. Finally the glueing link (black) is of standard length and is used where a permanent connection is to be made. A socket is also available which can be inserted into the user's specially made models if required. The locking links are used where multiple bonding is simulated.

In addition to the Research and Teaching Sets (5), CPK models can be obtained in various other kits for protein models (3), nucleic acids (2), steroids (1), new organic sets (2) and new kits for carbohydrate, polymer and surfactant molecular models. Moreover, it is possible to save time by purchasing preassembled structures. These cover a range of amino-acids and polypeptides, purine, pyrimidine units and DNA constituents such as the two base pairs, deoxyribose phosphate and a ribose phosphate chain. Alternatively, assembled DNA helices may be purchased.

These are excellent precision models, like the Courtauld ones, but with a more specialised application [10].

### 2.3.3 Scale Atoms (figure 2.9).

These are produced by Incentive Lärosystem, Stockholm, in four kits. The

Introductory Student Kit (I) is for secondary school chemistry up to O-level, the Advanced Student Kit (II) covers up to University entrance level and the Teacher and Research Kit (III) is for University use. In addition, there is a class-room kit consisting of sixteen small sets of models for very elementary teaching.

An atom model is a moulded, hollow sphere of strong, light, self-coloured nylon, consisting of a bowl and cap. The cap is of such a size that the distance from the centre of the sphere to the centre of the plane where the cap and bowl meet is approximately proportional to the covalent radius. The models are not accurately to scale because the spheres representing carbon, oxygen, nitrogen, sulphur and chlorine all have a radius of 17 mm (equivalent to a compromise van der Waals' radius of 0.17 nm) and the radius of the hydrogen atom model is 12 mm (≡0.12 nm).

Fig. 2.9 – Scale Atoms: (a) sulphuric acid, (b) neopentane, (c) *n*-pentane, (d) tyrosine.

The peg inside a bowl can be pushed into a hole on the outer surface of another bowl. A whole series of bowls linked in this way can form a molecular model, though once all the required atom models are joined together any with exposed inner surfaces must be completed by a cap. The simplest way of separating a bowl and cap is to prise them apart using another cap. Bowls are easily separated manually.

In set I the peg receptacles on carbon, oxygen, nitrogen and sulphur models are disposed tetrahedrally, except for an additional carbon model (distinguished by its grey colour from the black single bond species) which has receptacles arranged to receive pegs in either a trigonal or linear arrangement, so simulating double or triple bond carbon. There is also an additional oxygen model (orange, as opposed to the red of the single bond oxygen model) which represents carbonyl oxygen and which only has one receptacle, in the cap. The chlorine sphere, too, has only one receptacle. The multiple bond carbon atom model has square pegs rather than round ones so that it cannot be used by mistake instead of the single bond carbon atom, and also rotation about the link is precluded. However this device has not been used in the double bond oxygen model.

In an assembled molecular model every sphere is partly covered by one or more other spheres and the general appearance is very similar to that of previously discussed models in which segments of spheres are absent. As a result the outer surface of the Scale Atoms model gives a reasonable impression of the van der Waals' envelope by using an entirely different technique. Take the simplest case of a chlorine molecule, for example. A chlorine bowl is joined to a chlorine atom model by inserting the peg of one into the cap of the other so that the hollow interior of the incomplete sphere is covered.

Sets II and III, with which I is compatible, have several more different types of atom model and III has over three hundred components with which large assemblies can be constructed. Once again the set is particularly geared to the construction of molecular models representing biologically important species, up to and including DNA. Locking pins for strengthening such assembles are provided in kits II and III.

Kit II has a round peg version of the grey multiple bond carbon model for situations where free rotation is possible, for example about the C–C bond when a carboxyl group is attached to an aromatic ring. There is also a two-peg, wedge-shaped carbon model which facilitates the assembly of condensed ring systems. A square peg multiple bond nitrogen species with linear or trigonal symmetry is included and there are two other atomic species representing tetrahedral phosphorus (fawn coloured) and bromine (brown).

Kit III has all the above types and planar nitrogen with a round peg as well. It also has wedge-shaped nitrogen and oxygen, each with two pegs for polycyclic systems, and so-called 'universal' carbon and nitrogen. These have peg receptacles specially designed so that the atom models can form part of unsaturated five-membered rings.

None of the sets has any type of metal atom model.

A booklet by Larsson, 'Chemistry with Scale Atoms' [11], may be purchased which sets out a teaching programme to be used in conjunction with the elementary set (I). The intention is to provide the student with a means for self-tuition, but the book is very uneven in that parts seem to be vastly over-simplified while others are quite tough for beginners. A competent teacher can do without the book but should not be deterred from obtaining some of the models, which are remarkably cheap for space-filling models, very durable and easily assembled and dismantled. There is loss of precision, relative to the more sophisticated (and expensive) models discussed above, but this is of little importance at an elementary level. On the other hand, even at this stage much is gained by giving a student an approximate idea of the spatial characteristics of a molecule.

Instructions are included in both kits I and II but recently Complete Study Sets have become available. These consist of a student's model kit and manual, and a teacher's model kit and guide plus overhead transparencies.

### 2.3.4 Effex Stereo Molecular Models

Instead of a partial hollow sphere, an indented sphere with a moulded spigot is employed in the Effex Stereo Molecular Models System. The spigot, which is at the centre of the indentation, is pushed into a drilled hole on another sphere which may or may not be indented. Hence part of the rounded surface of the second sphere is covered and the effect is similar to that obtained using Scale Atoms.

A special feature of the Effex models is their large size, which makes them suitable for demonstration purposes. The spheres are of 2.5 and 3.0 inch diameter and are made of high-density polythene. They are colour coded according the the usual convention (Table 1.2).

Rotation is possible where single bonds are portrayed. Multiple bonds between carbon atoms are represented by permanently joined units.

Three kits of preassembled models designed as Teachers' Demonstration Sets at junior, senior and Technical College levels are available and there is also a 'Senior Organic Set' for 'O' and 'A' level. There are over thirty individual preassembled models covering a range of inorganic and organic structures.

In addition, kits of components for the user to assemble may be purchased. These are, 'Inorganic', 'Small Organic', 'Organic and Inorganic' and 'Benzene and Derivatives'.

These models, which originated in Australia and New Zealand, are made in the UK by CSL. They are very reasonably priced, especially when the large sizes of the spheres is considered.

## 2.4 HOMEMADE SPACE-FILLING MOLECULAR MODELS

The division between manufactured and homemade models is in some cases a somewhat arbitrary one, but in this context it is interpreted as meaning that the user has to do a certain amount of work in preparing the models for assembly. Some manufacturers provide kits of components, for example spheres and connectors, but the spheres have to be shaped and the positions of the connectors decided upon. There are also a number of devices on the market (described in detail in Chapter 9) which facilitate fabrication of the atom models and/or provide assistance with dimensional or symmetry problems.

There are two reasons for using homemade models. The first is the obvious one of avoiding the relatively high cost of purchasing the manufactured models.

The other reason, which may sometimes be of overriding importance, is the need to prepare a highly specialised model which is not available commercially. Perhaps the first consideration is paramount to a teacher, not only because sufficient funds for the purchase of expensive models may not be available but because of the high mortality rate suffered by most kinds of equipment in the hands of youngsters as well. A simple, cheap model, possibly even made by the students themselves, and preferably from readily available material, is therefore desirable for these. However, the research worker may find that the particular type of model which he requires is unobtainable. This may easily happen, for example when a new compound has been synthesised and its structure and properties are being investigated. In such a situation the researcher has no alternative but to make a special model himself. Unfortunately, however much one may enjoy making models (and model-making is recommended for its therapeutic value) and however satisfactory the result, making models from simple, commonly available materials is often a very lengthy process requiring not only time but patience, and a certain amount of expertise which is acquired only with practice.

To attempt to describe many types of homemade model individually would result in a long and possibly tedious Section so the discussion is confined to three topics, materials, connectors and kits of parts. It is hoped that these, together with the references at the end of the Chapter, will provide the would-be model-maker with sufficient information to set him on his way.

### 2.4.1 Materials

Model-making has become very much easier since plastics came upon the scene. Before these versatile materials became readily available a number of less tractable materials were used. Perhaps the most popular of these were wood, which had to be carved, or cork, which had to be cut. However papier-maché, which could be cut or sawn when dry, has been used, also clay and plaster of Paris, which could be moulded. Hard maple spheres with connector sockets arranged to give either the symmetry of a square antiprism or that of a trigonal dodecahedron were designed and made by Homeier and Larsen [12] for use with FHT models for the assembly of 8-co-ordinate metal complexes, no other suitable model being avaialble. Kellett and Martin [13] duplicated existing models by making them from silicone rubber or latex moulds. They filled these with surgical plaster and painted the product with water colours, the colour being preserved by coating with a spray plastic. New prototypes can be duplicated and also groups like half a benzene ring. The pieces can be glued together when dry.

However, because plastics can nowadays be more or less tailor-made to suit requirements, there seems to be little point in using less easily fabricated materials. Expanded polystyrene is the material of choice for most people as, according to its density, it can be cut with a knife, sawn with a fine hacksaw or

shaped using a hot wire cutter which melts its way through leaving a smooth surface. High density polystyrene is usually the most suitable because it can be sanded if necessary, painted with an ordinary household gloss paint and penetrated easily by some of the commonly used connector devices. Hard, dense models can be made from epoxy or polyester resins. Hoover and Shriver [14] used this material to duplicate wooden prototypes by making latex moulds of them and then filling these with the resin to which a mineral pigment or pharmaceutical dye had already been added. After casting, the surfaces were ground smooth and holes drilled for connectors. These models were used to supplement the Catalin covalent models where they did not meet specific requirements. A material with quite different properties is polyurethane foam, which is soft and spongy. This has been used by Doré [15] who cut spherical segments with radii proportional to the van der Waals' radii and glued them on to a wire framework in such a way that adjacent spheres were compressed so that the distance between their centres was roughly proportional to the sum of the covalent radii of the atoms which they represented.

### 2.4.2 Connectors

Of course, it is not essential to use any kind of connector because the atom models can simply be glued together using an appropriate adhesive. There are various proprietary brands of polystyrene cement but a reasonably satisfactory one is made by dissolving polystyrene scraps in toluene until a viscous fluid is obtained. It can then be applied with a glass rod. Polyurethane can be glued with Evostick or any similar adhesive. A versatile, 'reversible' adhesive known as 'Blu-Tack' can be obtained from stationers. This plasticine-like material is moderately priced and very economical in practice because it can be used over and over again, being virtually indestructable. A small blob placed on one of the surfaces to be joined will adhere to both surfaces when these are pressed together. It can be peeled off and re-used when required. Although a molecular model assembled with Blu-Tack will not fall apart unless subjected to really rough treatment, the atom models can easily be separated deliberately.

Perhaps the easiest connectors to use with polystyrene are metal or wood double-pointed dowels. They are usually sharp enough to pierce the polystyrene and short enough for the two atom models to be brought into contact. Alternatives are orange-sticks, and wooden or plastic cocktail sticks, all of which can be shortened and sharpened. The last tend to be rather brittle but have the advantage of being more flexible than wood. Ordinary wire can also be used but it may be necessary to heat it to get sufficient penetration into dense polystyrene. Unless care is taken, heating the wire can sometimes result in too large a hole being produced. For soft plastics, like polyurethane, flexible pipe-cleaners are suitable. Connector sockets must be drilled in the hard resins and then tubular rubber pegs or tightly fitting wooden or plastic pegs used. The most elegant way of connecting homemade models is by means of press-studs which can be purchased from manufacturers who use them in their own atom models. These are usually either glued into sockets on the models or screwed into position. Press-studs are the most satisfactory device when it is necessary to dismantle and re-assemble a model because this can be done many times without causing damage. With other kinds of connector it is often difficult or impossible to take the assembly apart and rebuild it.

### 2.4.3 Kits of Components

Although it has not appeared in recent catalogues, the Edmund Closed Molecular Model Kit may still be available. It consists of one hundred moulded polystyrene spheres of eight different sizes varying from 0.75 to 2.83 inch diameter. There are eighty one-inch and twenty one-and-a-half inch metal dowel screws for use as connectors. Red and black marker pins are used to represent electrons. The spheres must be cut to give the appropriate ratio of van der Walls' to covalent radius and then coloured. No means is provided for determining connector positions so as to produce the correct angles in a molecular model.

The Griffin Polyzote Spheres are available in six sizes ranging from 0.5 to 2 inch diameter. They can be cut with a sharp blade or sawn with a small hack-saw and then painted. No connectors are supplied but those described above for use with polystyrene are suitable. An instruction booklet is provided together with a paper protractor template for glueing on to card or wood. This greatly assists in the identification of 'bonding' sites with the appropriate angle. A scale of 0.5 inch ≡ 0.1 nm is recommended.

The Sargent-Welch Crystal Lattice and Molecular Models Kit consists of fifty-four polyurethane foam spheres in eight different colours and two different sizes. As the soft polyurethane foam is compressible it is not necessary to cut the spheres because they can be squashed together to represent covalently bonded species while their outer (uncompressed) surfaces represent the van der Waals' envelope. They are joined by metal rod connectors. The spheres are not permanently deformed by compression and holes from which connectors have been withdrawn will close up. They may therefore be used many times, not only as space-filling models but also as crystal structure models where the spheres represent ions and are in tangential contact (see next Chapter). They may be washed with soap and water and will retain their colour indefinitely.

There are many other sets of components. The three examples given above are only intended to give some indication of the types available.

Data on most of the models described in this Chapter are summarised in Tables 2.1 and 2.2. Price codes are only approximate, being an estimate of the outlay required for an adequate number of components which, where appropriate, is the cost of a kit of models. Sources of materials for homemade models are given in the next Chapter.

Construction devices and techniques for the preparation of homemade models are discussed in Chapter 9.

## REFERENCES

[1] Uses; some examples:
(a) Edward, J. T. *J. Chem. Educ.*, **47**, 261, (1970).
(b) Hendrickson, H. S. and Srere, P. A. *J. Chem. Educ.*, **45**, 539, (1968).
(c) Landé, S. *J. Chem. Educ.*, **45**, 587, (1968).
(d) Macnicol, D. D. and Wilson, F. B., *Chem. Commun.*, 1971, 787.
(e) Parfitt, G. D. *Ann. Rep. Chem. Soc.*, **64**, 152 and 160, (1967).
(f) Tsuge, S., Leary, J. J. and Isenhour, T. L., *J. Chem. Educ.*, **51**, 226, (1974).

[2] Stuart-Briegleb models; original publications:
(a) Briegleb, G. *Fortschr. Chem. Forsch.*, **1**, 642, (1950).
(b) Briegleb, G., in *Methoden der organischen Chemie*, Houben-Weyl, Stuttgart, (1955).
(c) Stuart, H. A., *Z. Phys. Chem. (B)*, **27**, 350, (1934).
(d) Stuart, H. A. *Die Struktur des freien Moleküls*, Springer, Berlin (1952).

[3] Godfrey, J. C. *J. Chem. Educ.*, **36**, 140, (1959).

[4] Hartley, G. S. and Robinson, C. *Trans. Far. Soc.*, **48**, 847, (1952).

[5] Robinson, C. *Courtauld Atomic Models, A Descriptive Textbook*, Griffin and George, (1966).

[6] Bent bonds:
(a) Bak, B. and Led, J. J. *J. Mol. Structure*, **3**, 379, (1969).
(b) Coulson, C. A. and Moffitt, W. E., *Phil. Mag.*, **40**, 1, (1949).
(c) Gassman, P. G., *Chem. Commun.*, 1967, 793.

[7] Corey, R. B. and Pauling, L. *Rev. Sci. Instr.*, **24**, 621, (1953).

[8] Koltun, W. L., *Biopolymers*, **3**, 665, (1965).

[9] Boyd, D. B., *J. Chem. Educ.*, **53**, 483, (1976).

[10] Harte, R. A., *Molecules in Three Dimensions: A Guide to the Construction of Models of Biochemically Interesting Compounds with CPK Models.*, 20-page booklet from the Ealing Corporation or its subsidiaries.

[11] Larsson, G., *Chemistry with Scale Atoms*, Heinemann (1970).

[12] Homeier, E. and Larsen, E. M., *J. Chem. Educ.*, **43**, 376 (1966).

[13] Kellett, J. C. and Martin, A. N. *J. Chem. Educ.*, **43**, 374 (1966).

[14] Hoover, W. G. and Shriver, D. *J. Chem. Educ.*, **38**, 295 (1961).

[15] Doré, C. F., *Educ. in Chem.*, **2**, 192, (1965).

# 3

# Space-filling crystal structure models

## 3.1 INTRODUCTION

This Chapter is chiefly concerned with models assembled from spheres in tangential contact representing crystal structures of ionic compounds. When the radii of the spheres used in the model are proportional to the radii of the ions represented, then a space-filling model is produced. Similar limitations apply here as are discussed in the preceding Chapter with respect to covalent bonding. A simple theoretical model of the ionic bond envisages ions as hard, incompressible spheres bearing a uniform charge on their surfaces. This enables them to be attracted to, or repelled by, other ions with unlike, or like, charges respectively. These forces are regarded as being exclusively electrostatic in nature and no account is taken of polarisation effects or covalent contributions. The inadequacies of this elementary theory are reflected in the crystal structure models used to illustrate it.

As with molecules, however, ionic crystal structures are most accurately portrayed with the aid of space-filling models, although the absence of motion and (usually) of defects again precludes a more realistic presentation. For precise work, spheres can be obtained whose radii are accurately proportional to the literature values for the ionic radii. The relative sizes of the spheres are then correct in the assembled model, providing that the application is appropriate of course. For example, ionic size changes with co-ordination number because it is determined by an equilibrium resulting from the opposing forces of attraction and repulsion between an ion and its neighbours. Therefore precision would be decreased if a sphere accurately scaled for six-co-ordination were to be used in an eight- or four-co-ordinate environment.

As with molecular space-filling models, much detail is obscured in assemblies of spheres in contact and the immediate environment of a given sphere (unless it is at the surface) is often only discernible with difficulty. Modified molecular models such as Courtauld and C.P.K. achieve a degree of 'transparency' by imparting some of the internuclear distance to the connectors. Employing such a device with spherical models would convert them into ball-and-spoke models (see Chapter 4) which fulfil different requirements on the whole. However, models built from spheres are particularly good at indicating overall symmetry patterns in three-dimensional arrays and are used extensively to demonstrate close-packing sequences and structures derived from them. Both the existence of tetrahedral and octahedral holes and their sizes relative to the close-packed spheres can also be shown. Discussion of these topics often involves a particular crystal plane, for example the (111) plane of a cubic close-packed sequence. If the spheres are of opaque material then the plane under discussion must be at the surface. On the other hand, if transparent spheres are used either these may be of a different colour from the rest or they may be marked with a spot of paint. In this way particular features of interest may be distinguished by looking through the assembly.

Most of the applications of this type of model which have been mentioned so far have been concerned with teaching, but assemblies of spheres are also useful in the study of surface reactions of ionic crystals at a research level. Speculation on the mechanism of gas-solid interactions is greatly facilitated when a stereochemical relationship between sorbent and sorbate can be visualised. Again, such features as the accessibility of active sites to reactant molecules and the feasibility of 'bridging' reactions spanning two sites are much more readily appreciated when both the solid surface and the gas molecules are represented by space-filling models on the same scale.

Models representing molecules are sometimes also made from complete spheres, in which case the radii of the spheres are usually proportional to the *covalent* radii [1]. These are not space-filling models because representations of the electron density on the non-bonding surfaces are lacking. Hence the limit to closeness of approach of another molecule or atom is not indicated as it is in true space-filling molecular models, where the van der Waals' envelope is defined. Spheres can also be used simply to identify atomic positions when neither relative ionic size nor the non-bonding surfaces need to be indicated and space-filling characteristics may be neglected.

## 3.2 ANCHORING SYSTEMS

Various anchoring systems can be used to hold spheres in tangential contact in suitable positions for depicting ionic crystal structures, and these are now discussed.

### 3.2.1 Glue or connectors

These are the commonest means of holding spheres in contact. Materials like expanded polystyrene, polyurethane or cork can be penetrated by the type of connector described in the previous Chapter and the strength of the assembly can be reinforced by glueing them in position. Spheres made of hard plastics like phenolic resins, acrylic, nylon or polyethylene cannot be penetrated but they can be drilled if they are not too brittle, in which case connectors made of rubber, wood or plastic may be used. Hollow spheres must be glued directly to each other or placed in some kind of container (see next Section). Clearly, the adhesive used should be appropriate for the material used to make the spheres but proprietary brands such as Evostick and Bostik are suitable for most plastics while the homemade glue described in Chapter 2 can be used for expanded polystyrene. A heated metal rod will bore a hole right through spheres of some plastics. They can then be threaded on rod, flexible wire, string or even elastic [2]. In the last case, a chain of spheres is drawn up into a helix when the elastic is tightened – a nice mechanical analogy. A very convenient, but inevitably more expensive technique for joining polystyrene spheres is to use an electric glue gun. This dispenses a suitable amount of hot melt adhesive accurately sited so that many spheres can be permanently joined in minutes and large assemblies built very rapidly. With most glues, on the other hand, time must be allowed for hardening or drying, which means that only small sections of an assembly can be built at any one time because the spheres have to be supported mechanically.

Where it is required that spheres-in-contact models should be assembled and dismantled many times, as when they are made available to a group of students, a 'reversible' adhesive is needed. I have found 'Blu-Tack' to be ideal for this purpose. To join polystyrene spheres all that is necessary is to place a blob of 'Blu-Tack' on one and press it to the other. While an assembly remains intact indefinitely and quite a large one is mechanically stable, it is easily taken apart when necessary. The 'Blu-Tack' can be peeled off and re-used. It is odourless, non-toxic and virtually indestructable. Most stationers supply it at a moderate price.

One application of 'Blu-Tack' with which I have had experience is in the construction of models of hexagonal and cubic close-packed structures. Two interpenetrating hexagonal prisms demonstrating the ABAB--- sequence can easily be constructed using polystyrene spheres and 'Blu-Tack'. The technique for construction of a face-centred cube is to take two sets of six spheres and join each set in a planar 3-2-1 arrangement. The two sets are then juxtaposed in a staggered arrangement and joined by 'Blu-Tack'. An extra sphere is placed on top of each six-layer so that it rests on the hole formed by the row containing two spheres and the central sphere of the row containing three. The face-centred cube is then completed.

It is possible to purchase ready-made assemblies in which the spheres are already permanently glued together (Figure 3.1). Science Related Materials Inc. produces radius ratio models in which the opaque central sphere representing the cation is surrounded by larger, hollow transparent spheres representing the anions. Four models illustrate trigonal, tetrahedral, octahedral and cubic co-ordination. In addition ten assembled crystal structure models, including some of the most important, and the two close-packing sequences are also available. The larger spheres are again transparent and the smaller opaque so that the positions of the latter are readily apparent. The relative sizes are correct and the whole assembly suitable for classroom demonstration. They are reasonably robust and not too expensive when labour costs are taken into account. Very similar models, including a range of assembled structures, may also be obtained from Estruteral. LeMont Scientific Inc. also market preassembled models of ionic crystal structures and of the two kinds of close-packing. Cork or plastic spheres are used which are colour-coded and of the correct relative sizes. Sixty-six assemblies are given in a recent catalogue and, like the SRM models, these are attractive, rugged, suitable for teaching and not excessively expensive. The component spheres may be bought unassembled from both these manufacturers. Many variants are possible with the SASM models, which are of French manufacture. This is because the conventional Stuart models (Chapter 2) can be used with long connectors, thus converting them effectively to ball-and-spoke models (Chapter 4). Moreover, crystal structures built from painted or lacquered box-wood spheres can be assembled using either very short plastic connectors, which bring the spheres into contact, or long ones to give ball-and-spoke models. These can all be obtained ready-assembled and the current catalogue lists nearly eighty inorganic structures of the tangent-sphere type which can be purchased. Others

Fig. 3.1 – SRM: assembled sodium chloride.

can be produced on request. The sizes of the spheres are proportional to the species which they represent and small, medium and large scales are available. Models of organic structures in which the atoms are represented by complete spheres can also be supplied already assembled and there is a wide range of options for the purchase of kits of components of various kinds.

The 'Molymod' system works on similar principles. With very short links approximations to space-filling models are obtained, but with longer links they are essentially of the ball-and-spoke type. These models can be obtained as components, as kits, or already assembled. They are discussed in more detail in Chapter 4. Other suppliers of spheres are given in Table 3.1. Some produce kits of components which include spheres, connectors and/or adhesive, instructions and (usually) construction aids of some kind. Spheres are occasionally self-coloured but it is often cheaper to buy them uncoloured and then to paint them. A simple technique with polystyrene is to pierce the sphere with a cocktail stick, dip it in the paint, drain off the excess and then insert the other end of the stick into a flat piece of polystyrene so that the sphere is supported vertically. Transparent spheres can be painted with transparent paints if necessary.

The teaching of structural inorganic chemistry at all levels is greatly facilitated by the use of 'spheres-in-contact' models. Judging by the numerous articles appearing in the major chemical education journals, the most popular topic demonstrated with these at a fairly elementary level is close-packing. A simple way of showing the difference between the two packing sequences is to make several layers (a minimum of four) by sticking at least twenty spheres of equal size together and then stacking either in an ABAB sequence (hexagonal close-packing) or in an ABCA sequence (cubic close-packing). Having put one layer over another it is easy to show that the third layer can fit into one of two sets of interstices by moving it about (Figure 3.2). The third layer may repeat the first layer or be different from it (ABA or ABC). If two layers are stacked on an overhead projector then the bright spots on the screen indicate the positions of octahedral holes. When placed in the octahedral holes, small spheres of suitable size black out the bright spots. This is a useful technique for showing partial filling of octahedral holes between layers, for example the two-thirds occupancy in chromium III chloride. A model which is complementary to the one just described consists of two sets of three spheres glued together, a layer in which one sphere is surrounded by six others and one on its own (Figure 3.3). An arrangement consisting of a set of three with a fourth placed over them has a tetrahedral hole at its centre. Two sets of three, one above the other, in a staggered configuration have an octahedral hole at the centre. Using spheres of appropriate size it can easily be shown that a very small sphere placed in a tetrahedral hole will distort the arrangement while a substantially larger one can be accommodated in an octahedral hole without distortion. The seven-sphere layer shows that a single sphere in a close-packed arrangement is in contact with six

**Table 3.1**
Spheres-in-contact models

| Type | Supplier* | Features of interest/ accessories | Price code (Table 1.4) |
|---|---|---|---|
| assembled | Estrutural LeMont, SASM, Science related Materials (SRM), Spiring Enterprises (Molymod) | colour-coded, size ratios correct where relevant | Estrutural a,b,c,d; LeMont a,b,c; SASM a,b,c,d; SRM a,b; Spiring a. |
| kits of components | (a) Griffin and George | (a) paper protractor template, instructions. | Griffin, a. |
| | (b) Macalaster Scientific | (b) poppit beads for [4] and [6] holes. | Macalaster, a. |
| | (c) Morris Laboratory Instruments | (c) 'Crystals Kit' includes some crystals and chemicals for growing crystals. | Morris, a,b. |
| | (d) Sargent-Welch Scientific | (d) compressible polyurethane spheres also suitable for molecular space-filling models, storage box and instructions. | Sargent, a,b. |
| | (e) SASM | (e) painted or lacquered wooden spheres, plastic connectors. | SASM, a,b. |
| | (f) Spiring Enterprises | (f) 'Molymod' system with shortest connectors. | Spiring, a. |
| spheres alone: | | | |
| (a) cork | LeMont, SRM | wide range of coloured cork spheres. | LeMont, a; SRM, a. |
| (b) polystyrene | Philip Harris, Plasteel, SRM | Plasteel markets a colour kit and coloured pipe-cleaners. SRM supplies a connector package, colour kit, glue gun and 'Angstrom ruler'. | Harris, a; Plasteel, a,b; SRM, a,b,c.‡ |
| (c) transparent | Ace plastic, LeMont, Oxford University Press (Wells), SRM | Ace and LeMont spheres are solid, Oxford and SRM hollow. Included with Oxford spheres are 15 solid acrylic smaller spheres and cement. | Ace, a; LeMont, a; Oxford, a; SRM, a,b,c. |
| (d) solid cellulose acetate | Ace, SRM | | Ace, a; SRM, a. |
| spheres with base: | Catalin, Estrutural and Griffin and George, Leybold, SRM | | Catalin, b.c; Estrutural, a,b; Griffin, a; Leybold, b,c; SRM demonstration b,c; SRM student, a. |

*Full addresses of suppliers are given in Appendix 1.
‡A wide range of prices reflects an extensive size range.

others in a single plane. If one set of three is placed below this layer and the other above it, it can be seen that the central sphere has twelve nearest neighbours. Moreover, if the two sets of three either side of the seven layer are in a staggered configuration with respect to each other, an extended array of this kind would be cubic close-packed (ABC). On the other hand, if the top set of three is aligned with the other set of three, the array would be hexagonal close-packed (ABA).

Fig. 3.2 – Layers of glued spheres demonstrating hexagonal close-packing.

A further use for stacked layers of spheres containing smaller spheres in the interstices is to demonstrate the difference between a three-dimensional and a layer crystal structure. For example, the nickel arsenide and cadmium iodide structures can both be described in terms of hexagonal close-packing of arsenide or iodide with the metal ion in octahedral holes. In the nickel arsenide structure all the octahedral holes are filled and in cadmium iodide only half. However, the latter is a layer structure and so all the holes are filled between alternate layers. Other structures which can be demonstrated in this way include the three-dimensional corundum structure. Here the oxide ions are hexagonal close-packed with aluminium ions distributed as evenly as possible in two-thirds

Fig. 3.3 – Layers of three, seven, three and one spheres used to demonstrate close-packing sequences and tetrahedral and octahedral holes.

of the octahedral holes. The chromium (III) chloride layer lattice consists of chloride ions in a cubic close-packed arrangement with one third of the octahedral sites overall being filled. However, because this is a layer lattice the chromium ions actually fill two-thirds of the sites between every other layer. These assemblies are best built in front of the students and are preferably used in conjunction with ball-and-spoke or framework models (see Chapters 4 and 5 respectively). Alternatively, if transparent spheres are used for the close-packed framework and coloured opaque spheres for the cations, the positions of the latter are easily seen. The models help to explain the distinction between three-dimensional and layer structures in terms of the nature of the bonding. Using them it may be clearly recognised semi-quantitatively that because the anions are much larger the distance from the centre of a cadmium ion to the centre of an iodide ion is less than that between the centres of two iodide ions. While primary bonds (and hence shorter internuclear distances) exist between cadmium and iodide, only van der Waals' long range forces exist between the anions and therefore these are further apart. On the other hand, the aluminium ions in corundum are uniformly distributed in a three-dimensional structure. Although some lattice distortion is caused because one third of the cation sites are empty, there is no line or plane of weakness in the structure.

Colour can be used to good effect with tangent sphere models. Some years ago I made a model for illustrating the differences and similarities of the hexagonal and rhombohedral crystal systems in which the spheres comprising the rhomb are coloured and the others are not (Figure 3.4). With this model the 6-fold symmetry of the hexagonal system, the 3-fold symmetry of the rhomb and their relationship are readily seen. Most diagrams, on the other hand, are confusing.

Fig. 3.4 – The related hexagonal and rhombohedral crystal systems.

Major contributions to the literature on this type of model have been made by Sanderson [3] and Wells [4]. Sanderson's book is a useful guide for teachers at both secondary and tertiary levels and provides a wealth of chemical and practical information giving step-by-step directions for the assembly of a variety of models. The majority of these are tangent sphere models including true space-filling models which are our chief concern in this Chapter. A number of 'lattice' models of common crystal structures are also illustrated. Here the spheres are all the same size because they represent the positions of the ions in a lattice rather than the ions themselves. Cation and anion positions are distinguished by using different coloured spheres. Sanderson also deals with ball-and-spoke models and molecular models of spheres in contact in which the size of the sphere is proportional to the covalent radius.

A system for teaching advanced structural inorganic chemistry with the aid of models is admirably presented in Wells' book. Although the book is intended to be used in conjunction with a set of models designed by Wells (Table 3.1), other similar models can be substituted. Various types are used in addition to tangent spheres, and these are mentioned in the appropriate Chapters. The sections on spherical models deal with important binary structures and some complex oxides and halides as well as packing models and radius ratios. The spherical components consist of two hundred hollow, transparent plastic spheres of unit radius relative to the others (actually 1.5 inch diameter), half of them coloured. There are also smaller, coloured solid spheres of relative radii 0.225 and 0.414 which fit respectively into tetrahedral and octahedral holes formed by the larger spheres. There is also a single sphere of relative radius 0.732 which forms the central sphere in a cubic arrangement. Close packed $XO_3$ structures such as perovskite can be built using coloured spheres to represent X, in this case calcium, with colourless ones representing oxygen in a ratio of one to three. This book is more specialised and more advanced than Sanderson's. It is warmly recommended to teachers of structural inorganic chemistry at tertiary level.

References to other literature dealing with this type of model are given at the end of the Chapter.

### 3.2.2 Assemblies with Additional Mechanical Support

The simplest way of supporting spheres is to put them in a box. If the box has transparent sides it can become a useful teaching tool, particularly if transparent spheres are also used (Figure 3.5). Shaking the box can show that the spheres tend, with time, to order themselves so as to approximate more and more nearly to a close-packed arrangement. This can be seen to be a more efficient use of space because, if sufficient spheres are used, the total apparent volume occupied by them clearly diminishes as they are shaken. 'Defects', represented by spheres of a slightly different size, can be seen to introduce distortions in the arrangement and suitably small spheres are observed to fall naturally into the interstices formed by the larger spheres, that is tetrahedral or octahedral holes. A similar simple but two-dimensional technique (similar in principle to a bubble-raft) is to float spheres on water. Here again 'defects' can be introduced by using spheres of a different kind and 'dislocations' induced [5]. Small, hollow, polypropylene spheres such as those used as a floating lid for water baths are suitable.

Fig. 3.5 – Transparent spheres in perspex box showing packing arrangements.

Instead of placing the spheres in a container they may be built up from a base, lateral support being supplied in a variety of ways. The Catalin 'Adjustable Base' permits quite precise adjustment of sphere positions in three dimensions (Figure 3.6). The rectangular base carries two side-rails along opposite edges. These are linked by a number of cross-rails which are attached to the side by nylon sleeves so enabling the cross-rails to move along the side rails. Each cross-rail holds a number of sleeves with sockets on their upper surfaces and into these rods are placed which are then held vertically. These can be moved along the cross-rails. Adjustment in the vertical direction is brought about by using polythene spacers of suitable length threaded on to the rods between the spheres. The base is available in several different sizes. The spheres are of solid, self-coloured plastic with holes drilled right through them so that they can be threaded on to the vertical spokes. Forty-one different spheres in the Catalin ionic series represent thirty-six elements while the scale is the same as that of the Catalin Covalent models, namely 10 mm $\equiv$ 0.1 nm (Chapter 2). The spheres are colour coded according to the position of the element from which the ion is derived in the Periodic Table. Spheres representing cations are opaque and anions transparent, except that ammonium and hydroxide are mottled. This type of model is particularly good at portraying crystal structures in which one of the sub-lattices is close-packed because the occupants of tetrahedral and octahedral holes can easily be inspected by lifting overlying spheres along the vertical spokes. They then drop back into their correct position when released. The model can be built so as to expose particular crystal planes. This is not only useful for teaching but it can sometimes help in the interpretation of surface reactions, especially if the Catalin Covalent molecules are used to make up a molecular model of the adsorbate. (Unfortunately the surface of a solid is likely to be much more complex than any crystal plane within its bulk under most laboratory conditions. However, the matching of models can be an intriguing exercise!) Another useful function of the Catalin Ionic Model is the portrayal of dislocations and other defects in, for example, a metal crystal. The model is so easily adjusted that it is relatively simple to indicate an edge or screw dislo-

Fig. 3.6 – Catalin adjustable base: caesium chloride.

cation with it. The adjustable base can also be used for the assembly of disordered structures.

Other manufacturers and individuals have produced similar devices to the Catalin one. Leybold market a simpler system in which the base is of steel and vertical rods, supported by magnetic feet, can be attached to it at any point. The spheres are threaded on to the vertical rods as in the Catalin model. The spheres supplied with the base plate and rods are all the same size, and are best used for 'lattice' models or for indicating symmetry properties [6] (Figure 3.7).

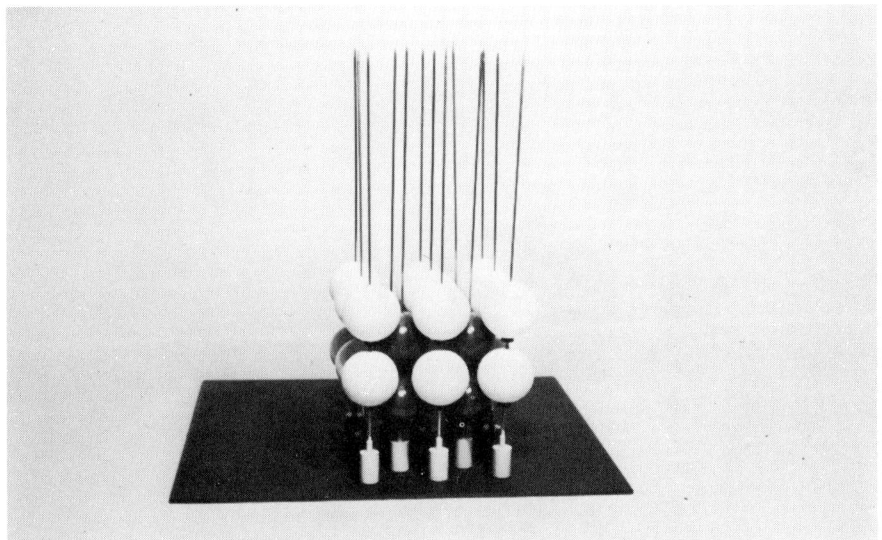

Fig. 3.7 – Leybold base: caesium chloride.

The Griffin jig for the sodium chloride crystal lattice consists of an aluminium alloy sheet with two edges bent down at right angles so that it is supported 20 mm off the bench or table. The sheet contains fifteen holes which can accommodate 50 mm polystyrene spheres representing chloride ions. The holes are arranged so that a face centred cubic lattice can be built up from the bottom layer of spheres. Sodium ions are represented by 25 mm spheres which fit into octahedral interstices formed by the larger spheres. The completed assembly consists of a cube with three large and two small spheres on each edge. Instructions are included in the kit.

The SRM 'Universal Crystal-Molecular Model' is also relatively simple. There are two versions. The demonstration size model has a base with vertical supports for five 'Plexiglass' clear, transparent, horizontal planes. These have holes punched in them into which appropriately sized spheres are placed. The geometry of the holes corresponds to atom (or ion) positions in common crystal planes and the spheres placed in them may or may not be in tangential contact. Lattice or packing models may be assembled. The coloured polystyrene spheres supplied are of different sizes. The student version has wooden sides into which five clear plastic 'planes' are slotted horizontally. The kit contains ten planes in all, five 'cubic planes' and five 'hexagonal planes'. Each has twenty-five holes punched in it to allow a variety of positions for the painted wooden spheres representing atoms (or ions). Once again both 'lattice' or 'packing' (spheres-in-contact) models can be constructed (Figure 3.8), and both kits include a storage box and instructions. Estrutural also supply a kit of forty models of this type with all accessories.

Fig. 3.8 – SRM universal student model: sodium chloride.

An effective teaching aid of a similar type has been designed by Onyszchuk [7] for student use. This consists of a platewood base in which holes are drilled on each side, one side forming a template for both the close-packing lattices (hcp and fcc) and the other for body-centred cubic (bcc). The former can also be used for explanation of the sodium chloride, zinc blende, fluorite and anti-fluorite lattices, although it is not possible to put spheres in tetrahedral holes. The latter can be used for the interpenetrating primitive cubic lattices of caesium chloride. The holes are colour-coded to assist identification of the different symmetries. To construct a particular model, vertical wooden rods (also coloured) are placed in the appropriate holes and polystyrene spheres which have been drilled right through the middle as in the Catalin models are threaded on to them. In this way the close-packed frameworks or interpenetrating simple cubic lattices are built up. The sodium chloride structure is produced by pressing smaller spheres into the octahedral holes and the caesium chloride structure by using spheres of the same size but differing colour in alternate layers.

A design by Robertson [8] is similar in principle in that plywood templates are used for the construction of caesium chloride and sodium chloride models. However the holes drilled are large enough for the bottom layer of spheres to rest in them while subsequent layers are supported by perspex sheets mounted vertically in grooves on the edges of the base to form a box with transparent sides. Appropriately painted polystyrene spheres are used, the same size (25.4 mm) for caesium and chloride but approximately half this (12.5 mm) for sodium. The caesium chloride model is easily constructed simply by placing a layer of caesium spheres at the bottom, then chloride and so on up to five layers. The spheres for the sodium chloride models are joined by pins in sets of three, $Na^+Cl^-Na^+$ and $Cl^-Na^+Cl^-$. The model can then be built up by placing these sets in the appropriate positions. The models are used by pupils studying for the Higher Grade in Chemistry of the Scottish Certificate of Education so they would be suitable for similar introductory courses in structural chemistry.

Both these and the Onyszchuk models are robust enough to be used many times over by students in schools and colleges.

Dynamic models of spheres in contact are discussed in Chapter 8, location of 'bond' positions in Chapter 4.5. Construction devices and techniques for the preparation of homemade models in general are considered in Chapter 9.

## REFERENCES

[1] Haworth, Sir Norman. In Haworth Memorial Lecture by M. Stacey, *Chem. Soc. Rev.*, **2**, No. 2, 145, (1973).

[2] Tetlow, K. S., *Educ. in Chem.*, **1**, 7, (1964).

[3] Sanderson, R. T., *Teaching Chemistry with Models,* Van Nostrand (1962).

[4] Wells, A. F., *Models in Structural Inorganic Chemistry,* Oxford: Clarendon (1970).

[5] Booth, N. *Educ. in Chem.*, **1**, 18, (1964).

[6] Kakabadse, G. J. and Theobald, D. W., *Educ. in Chem.*, **4**, 135, (1967).

[7] Onyszchuk, M. *Canad. Chem. Educ.*, **5**, (2), 6, (1970).

[8] Robertson, R. M., *School Sci. Rev.*, **53**, No. 182, 147, September 1971.

**General Bibliography**

1. Bassow, H. *Construction and Use of Atomic and Molecular Models,* Pergamon (1968). (Tangent sphere and other models, A-level/first year university).
2. Dabby, R. E., *Making Crystal Models,* Pergamon (1969). (Tangent sphere and ball-and-spoke models, A-level).
3. *Nuffield Chemistry: Handbook for Teachers,* Longmans/Penguin, (1967). (Various types of model as teaching aids discussed, including tangent spheres, A-level mainly).

# 4

# Ball-and-spoke models

## 4.1 INTRODUCTION

A continuous spectrum of model types exists which ranges from true space-filling models (discussed in Chapters 2 and 3) to skeletal models (Chapter 5). Ball-and-spoke models comprise the large central section of this spectrum and the boundaries at either end are not sharp. Indeed, assignment of certain models to a particular category is somewhat arbitrary and, in a few cases, difficult to rationalise. At the space-filling/ball-and-spoke interface we have spheres or shapes derived from spheres joined by connectors so that they are in contact. Lengthen the connectors or insert them only a short distance into the spheres so that these are not in contact, and we have a ball-and-spoke model. At the ball-and-spoke/skeletal borderline, the emphasis has changed in a different direction. Now the spheres have almost disappeared and the spokes, indicating molecular or crystal geometry, are all-important.

Therefore we can no longer regard the spheres in ball-and-spoke models as representing atoms or ions, only the *positions* of these. In other words these are 'lattice' models when they are used to represent crystal structures. If they are appropriately scaled the distance from the centre of one sphere to that of another represents the internuclear distance. Space-filling models remain, *par excellence,* the best for indicating steric hindrance and so suggesting by inference possible reaction mechanisms. On the other hand skeletal models (see Chapter 5) indicate overall crystal or molecular geometry and, if precise, subtle conformation differences.

Ball-and-spoke models are very versatile. In their fashion they can represent any crystal or molecule. They are extremely popular and are used extensively by crystallographers, solid state physicists and mineralogists as well as by chemists. Although they have been used in research, they are mostly used for teaching purposes and find application at all levels of chemical education. Their popularity probably stems from three causes. In the first place, they are virtually self-perpetuating in that students who ultimately become teachers are introduced to this type of model at school and/or college and perhaps do not have much experience of other kinds. The natural tendency is for them to continue to use ball-and-spoke models with their own students. Secondly, reasonably satisfactory models are easy to make from commonly available materials which cost little money. Moreover, the chemical education literature contains much advice on how to assemble these models, while a considerable amount of general knowledge about them among teachers and technicians is passed on by word of mouth. These comments are in no way intended to denigrate ball-and-spoke models, which perform a useful function. They are simply an attempt to explain their almost exclusive use in some educational institutions.

Pedagogically their most valuable attribute is their 'transparency', and this constitutes the third reason (in my view) for their popularity. The structure is clearly very much more open than in space-filling models and many chemical features are more easily seen as a result. This applies particularly to co-ordination numbers, measurement of bond lengths and bond angles, recognition of elements of symmetry, identification of crystallographic planes, differences between isomers, allotropes and polymorphs and so on. Muertterties and Wright [1] give excellent examples of how ball-and-spoke models can be used to illustrate both the similarities between certain co-ordination geometries and the small amount of distortion necessary to change from one to another. Their review contains a number of perspective photographs of groups of ball-and-spoke models with differing arrangements about the central sphere. For example, the idealised square antiprism and tricapped trigonal prism are both shown and it is apparent that the increase in co-ordination number from eight to nine requires little distortion. Similarly, it can be seen that there is no great difference between the idealised bicapped square antiprism and bicapped trigonal dodecahedron, both of which exhibit ten-co-ordination. Several other examples of these types are given. The combined effects of habit, cost and convenience, together with specific qualities, all produce powerful motives for using these models. In the UK they seem to be particularly favoured by university inorganic chemistry departments, possibly due in part to the availability of high quality models which may be purchased, already assembled, direct from the manufacturers.

In the most precise ball-and-spoke models the distances between spheres are proportional to the appropriate bond lengths in the crystal or molecule and the bond angles are also accurately reproduced (within 1-2°). However, for many purposes approximately correct angles and spoke lengths will produce an adequate model. Where the spoke lengths are accurately scaled, inferences can be made about the bond type because internuclear distances corresponding to 0.4 nm or more may be regarded as nonbonding. In this way primary or hydrogen bonds may be distinguished from van der Waals' forces.

The spheres of a typical ball-and-spoke model are all the same size but some indication of the relative sizes of the atoms or ions can be obtained by inserting pieces of rod or tube into spare holes in the ball. Crystal Structures Ltd. use flexible plastic rods which are inserted into holes covering the whole surface of the sphere. The rods are cut to size so that the van der Waals' radius is indicated. This device is known as a 'hedgehog'. When a primary bond is simulated, the 'hedgehogs' interpenetrate and it is sometimes possible to gain some idea of polarisation effects.

A few manufacturers have substituted polyhedra for spheres in their 'ball-and-spoke' models, presumably to help the user identify the various drilling geometries. However the underlying principles of these models are the same as for the conventional types. An alternative approach is to use long connectors between space-filling atom models of the Stuart type, effectively converting the whole into a ball-and-spoke assembly.

Spheres are usually made of wood, rubber or plastics. Hard materials have to be drilled and some device is necessary to ensure that angles between the holes are correct. The most useful drilled sphere has twenty-six holes arranged to coincide with the symmetry axes of the cube and so providing the common bond angles. Many manufacturers supply appropriately drilled spheres but the amateur model-maker is still faced with the problem of identifying the bond locations on the spheres. Some techniques which have been described in the literature are given in Chapter 9 and in Section 5 of this Chapter.

The connectors can be rods, springs, sticks or tubes made of metal, wood or plastics, including those discussed in Chapters 2 and 3 such as pipecleaners and cocktail sticks. Rigid metal rods are the most suitable for precise, permanently assembled models but for ordinary purposes flexible spring connectors are very satisfactory and have several additional advantages over rigid spokes. For example, two or three can be used to portray multiple bonding. The effect is to shorten the distance between the spheres and, by the bowing outwards of the springs, to give some indication of the electron density distribution due to the $\pi$-bond. When a structure is strained, the springs bend in such a way that both the presence of strain is indicated and some strained structures can be built which would have been impossible with straight rigid spokes. A further use is the demonstration of vibratory modes in small molecules because models of these made with springs can be easily distorted.

The linkage mechanism with all these models must be strong and, when a model is not to remain permanently assembled, able to withstand constant use. Assembly should be quick and easy and not too hard on the hands. The fit of the spoke in the socket should be tight without being too difficult to insert. Recommendations for permanently assembled precision models have been published by the crystal structure model panel of the X-ray Analysis Group of the Institute of Physics [2]. Many manufacturers adhere to these, and they also provide a useful guide for the amateur model-maker (Table 4.1). The Institute of Physics colour code is given in Table 1.2.

Like all models, the ball-and-spoke variety have several major deficiencies, particularly when considered from a student's point of view. For example, an ionic structure is represented by spheres all the same size, which might be confused with ions. They are not in contact but are separated by rod or tube which

**Table 4.1**
Standardisation of crystal structure models.
Institute of Physics recommendations.

| | |
|---|---|
| Scales: | 1 cm ≡ 1 Å (10 mm ≡ 0.1 nm) for small models<br>2.5 cm ≡ 1 Å for general purposes (most frequently used)<br>5 cm ≡ 1 Å for large models required for lecture demonstrations |
| Precision: | for simple structures internuclear distances should be correctly scaled to within 0.1 Å (0.01 nm). |
| Materials: | should be durable and the dimensions should not change over an indefinite period of time. |
| Spheres: | should not exceed equivalent of 1 Å (0.1 nm) in diameter but are preferably somewhat smaller. For a scale of 2.5 cm ≡ 1 Å, spheres of 2 cm diameter give sufficiently 'open' model. Generally impracticable to use spheres of different sizes in one model but if this is done, sizes should be in correct sequence but not necessarily to scale. |
| Connectors: | must be held securely. May be any material or gauge if sufficiently strong. Should not bend readily and should be protected against corrosion. For 2.5 cm ≡ 1 Å scale, stainless steel wire 3/32 inch diameter, hard-drawn and polished is suitable. Connectors should only represent 'bonds' except that the unit cell may be outlined. |
| Content: | all models should include at least one unit cell and sufficient adjacent atoms to show full co-ordination of all atoms in cell. |
| Label: | should be durably engraved giving name of structure, chemical composition, scale, whether 'idealised', a literature reference for complex structures and colour code (optional). |
| Colour: | should be permanent and coded according to the scheme given in Table 1.2. |

might be supposed to simulate the ionic bond between them. All may be well if it is emphasised to the student that what is really portrayed by this model is an array of points in space. However, if the idea of a crystal structure persists then two further problems arise. First the fact that this model is 'perfect' and secondly that its components are stationary. Although these last difficulties apply to virtually all models except those dynamic models specifically designed to show defects (Chapter 8), they matter more with ball-and-spoke models because these are used so extensively in teaching. However, providing these points are clarified they remain easily constructed, relatively inexpensive and valuable educational aids which can be assembled with varying degrees of sophistication. They can be used to construct complex organic molecules such as DNA and extensive three-dimensional arrays of inorganic structures (Chapter 6) but it would be fair to say that other types of model are more often appropriate for these.

## 4.2 ASSEMBLED MODELS

### 4.2.1 Beevers Miniature Models (Figure 4.1).

While the overall scales for most commercially available assembled ball-and-spoke models are of the order of 10 mm ≡ 0.25 nm or larger, the scale for the Beevers' models is 10 mm ≡ 0.1 nm. These models are composed of colour-coded Perspex spheres 6.9 mm in diameter (4.9 mm for hydrogen) and of stainless steel rods, and are very attractive and relatively inexpensive. Forty-five individual models or groups such as the five regular solids are listed and others can be made to order if required. The range includes the majority of structures usually considered important in courses on inorganic crystal chemistry up to and including University level. Mineral structures of varying complexity are represented and several molecules, including an alpha helix. With such small models extended arrays can be built which sometimes reveal subtleties in the structures not apparent in more restricted assemblies. I have examined such extended arrays of sodium chloride, quartz and the molecular crystal of benzene, among others, and there is no doubt that educationally these models are very valuable. In the first case, the way in which summing of electrostatic interactions gives rise to the Madelung constant becomes much more credible. A new view of the quartz structure is given in Dr. Beevers' paper [3] and the juxtaposition of hydrogens in a cluster of adjacent molecules in the benzene crystal provides convincing evidence of the spatial requirements of $\pi$-bonds.

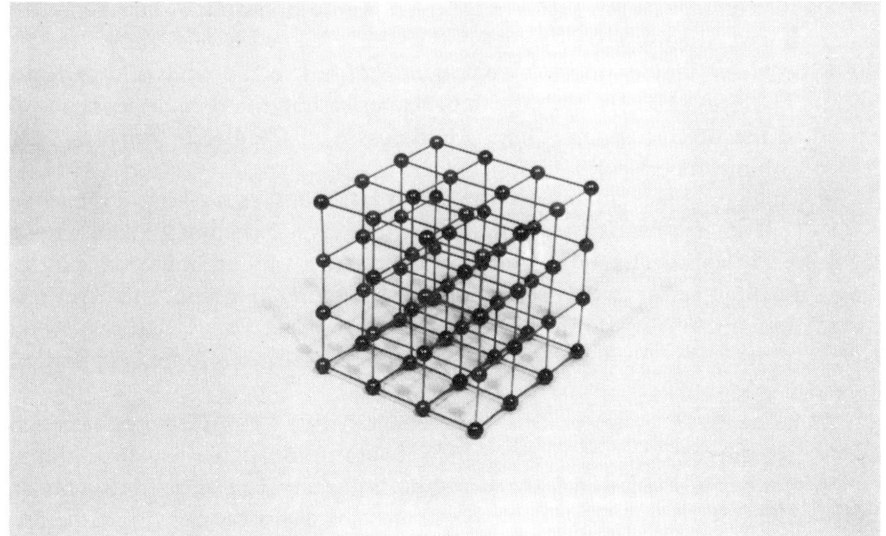

Fig. 4.1 – Beevers miniature assembled: sodium chloride.

The models are light and, because of their small volume, easier to store but they are inevitably less robust than larger models. However, the model can be kept in a Perspex case which may be purchased at the same time. Dimensions are accurate (according to Dr. Beevers they aim at an accuracy of 0.01 mm) and the bond angles are correct to within 0.5°. Attached to each model is a legend giving crystallographic and other data.

### 4.2.2 Crystal Structures Ltd. Permanent Models (Figure 4.2)

An enormous range of assembled models is available from this Company. A recent catalogue lists more than 450 prototypes. In addition, models usually can be made to order within a few weeks and a quotation can be given providing all structural data are supplied. Crystal Structures Ltd. have been an independent company since 1947 and examples of their deservedly popular 'P' series of accurate assembled models can be seen in many educational and research institutions in the UK. They are also marketed in the USA. (See Appendix 1 for distributors). The scale for the 'P' range models is 25 mm ≡ 0.1 nm and the painted wooden spheres with a diameter of 25 mm are permanently (hence 'P') joined by stainless steel spokes. They conform to all the Institute of Physics recommendations (Tables 1.2 and 4.1) except that the spheres are somewhat larger, at 25 mm as opposed to the recommended 20 mm. These are typical ball-and-spoke models in that all the balls are the same size but differences in radii (van der Waals' or primary bonding) in atoms or ions can be indicated by inserting plastic rods of appropriate colour and length into the balls. As previously mentioned, this device is known as a 'hedgehog'. Selection of individual models for comment from such a large range would be invidious but the manufacturers have themselves categorised them into elements and alloys, oxides and hydroxides, simple inorganic compounds, organic compounds and minerals (including clays). Models of large, biochemically important molecules are included in the organic group. These are discussed in Chapter 6, which is concerned with macromolecules.

Assembled models intended to be used particularly as teaching aids include a set of eleven models for teaching elementary crystal chemistry. Also among them are twelve silicate anion structures (mounted), various amino acids including a mounted set of twenty representing those found in insulin, and other sets for schools up to VIth form level. Sets demonstrating Bravais lattices, Miller indices, dislocations, Brillouin zones and reciprocal lattices also find application in teaching.

Some of the 'P' range models are available with 50 mm diameter balls and an overall scale of 50 mm ≡ 0.1 nm. These are useful for lecture-demonstrations.

A number of models, mainly of organic molecules, can be obtained with an overall scale of 10 mm ≡ 0.1 nm. These are designated the 'SP' (small permanent) series and they are usually mounted on a board on which are given crystallographic data, name and composition. A set of twenty amino acids are mounted together in this way. Because the spheres are 8 mm in diameter the overall appearance of the assembly is similar to a Beevers model and it has the same advantages, of lightness, easy storing, clarification of molecular positions in an extended crystal and so on. Spokes necessary for structural reasons but not representing bonds may be sleeved to mark them out. Spheres marking the corners of the unit cell are marked by white pins. In the larger models painted white dots are usually used.

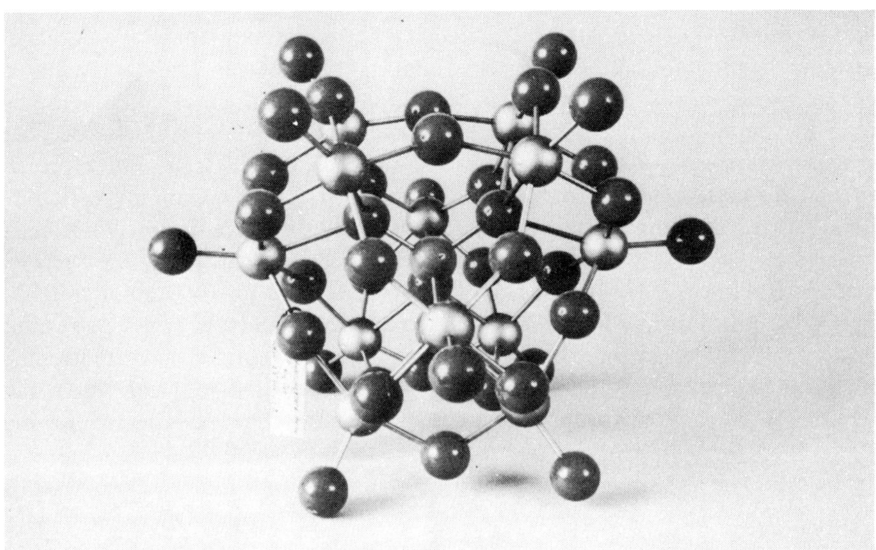

Fig. 4.2 – CSL assembled: 12-tungstophosphate anion ($PW_{12}O_{40}^{3-}$).

### 4.2.3 Leybold Crystal Structure Models

Leybold list about 100 assembled crystal structure models covering elements, essentially ionic compounds, silicates, organic compounds and alloys. Two sets of lattice models are available, one consisting of the fourteen Bravais (translation) lattices and the other being supplementary to it. A further set of 'important structure types' contains models of copper, magnesium, diamond, graphite, rocksalt, caesium chloride, wurtzite, calcite and the triclinic translation lattice. This is similar to, but not quite as useful as, the CSL Elementary Teaching Set which contains eleven models.

The size of sphere (25 mm diameter) and overall scale, 25 mm ≡ 0.1 nm, is the same as the CSL main 'P' range and the models conform to the Institute of Physics recommendations in all essentials.

American distributors of the Leybold assembled crystal structure models are given in Appendix 2.

### 4.2.4 SASM Assembled Models (Figure 4.3)

Although the actual number of structures represented (about 80) is very much smaller than for the CSL series, several different options in choice of scale, material or type of model permit many variations of the SASM assembled ball-and-spoke models. A good selection of inorganic crystal structure models is available covering elements, oxides, halides and compounds in which all the common anions (and some less common) are present. There are eighteen silicate models demonstrating the various types of anionic structure, also close-packed frameworks, metals, alloys and intermetallic compounds. A set of Bravais lattices can be obtained. Molecular models of co-ordination complexes, organo-metallic compounds and biologically important molecules can be purchased either with spheres to indicate atomic positions or with the type of atom model used in the space-filling series. These models are moderately precise because only six different spoke lengths are used for each overall scale. Therefore although they are approximately proportional to the most commonly occurring bond lengths encountered in organic chemistry the total accuracy will not be as high as in the most precise of those models described earlier in this Section. Three overall scales are possible, 25 mm ≡ 0.1 nm, 50 mm ≡ 0.1 nm and 100 mm ≡ 0.1 nm, but in addition the spheres can be obtained in various sizes, for example 10, 20 or 25 mm diameter. In fact, so many permutations are possible that it is not possible to describe them all in detail, added to which the manufacturer is prepared to supply any other variant on demand. The spheres are

Fig. 4.3 – SASM assembled: rutile ($TiO_2$).

of painted boxwood, used with plastic tubes in the several series designed for teaching, but with metal rods (which are scaled with greater accuracy) for the research series. Several basic sets are available and these can be augmented by complementary sets of a similar nature.

In combination with the SASM space-filling models (Chapter 2) the number of variations is very large indeed because several scales and sphere (or atom model) sizes are possible. Moderate accuracy is achieved in most cases but the research series are quite precise.

All the crystal and molecular structures can also be obtained in so-called 'compact' (or space-filling) assemblies.

### 4.2.5 Science Related Materials (SRM)

Pre-assembled models of biochemical species are supplied by these manufacturers but they are discussed with other macromolecules in Chapter 6.

### 4.2.6 Spiring Enterprises Ltd. (Molymod)

Twenty-four assembled models of specific structures are listed, most of them with either short or long links. Here is a good example of the problem posed at the beginning of this Chapter. The manufacturers describe the models with short links (bringing the atom models almost into contact) as space-filling but my own subjective assessment of them is that they are more like ball-and-spoke models, hence their inclusion in this Chapter rather than the preceding one (Fig. 4.9). Certainly, with long links they are unambiguously ball-and-spoke models (described as 'open-type' by the manufacturers). With the exception of the crystal structures of diamond, graphite, phosphorus, sodium chloride, zinc blende, calcite, rutile and ice, the models are all molecular, mostly of polymers and naturally-occurring macromolecules such as DNA, which is only available in the 'open-type' assembly. In addition to the individual structures there are assembled collections, for example of various hydrocarbons, alcohols, other aliphatic compounds, amino-acids and aromatics. These are only produced with short links.

The atom models are of solid, hard, coloured polystyrene with the holes for links moulded in. Around the link site is a planar region against which part of the link rests when two atom models are joined. The links are of slightly flexible polythene with a disc-shaped section near each end in the two longer links, and in the middle of the short link, which determines the extent of insertion of the link into the atom model. Further details of components are given in Section 4.4.

Scales are not mentioned in the manufacturers' literature but examination of the models themselves shows that, with short links, the scale is intermediate between those of the Beevers and CSL assemblies. With medium or long links it is similar to the corresponding Gallenhamp models (see Section 4.7 below).

## 4.3 SPECIFIC COMPONENT KITS

The CSL Z series comprises fourteen different kits each containing the appropriate number of suitably drilled and coloured spheres with slit spokes (see Section 4) for building a particular crystal structure. The structures include all those normally considered necessary for a modern VIth form chemistry course and a great deal would be learned by any student who successfully undertook the assembly of them. Assembly instructions are available separately. The scale is 25 mm ≡ 0.1 nm.

The Beevers brochure gives the number of large balls, small balls and rods required for the specific structures listed and the components can be purchased as separate items. Hence it would be possible to acquire specific kits for assembly of these miniature models.

SRM market thirteen kits (Figure 4.4) for assembly of the same crystal structures as the CSL Z series with the exception of iron. The spheres are of expanded polystyrene, colour-coded and with the bond positions marked ready to receive vinyl receptors into which the wooden spokes are inserted.

Fig. 4.4 – SRM specific kit: sodium chloride.

Use of the vinyl sleeve (which has to be cemented in) facilitates assembly and disassembly so that the kit can be used many times. These kits are known as the Crystal Construction Atoms Sets, abbreviated to CCA kits. They are useful teaching aids. Construction of what is believed to be a thoroughly known structure can be a salutary experience for the student, and sometimes for the teacher too!

Ten SRM demonstration-size kits of metal structures, common ionic crystal structures, diamond and graphite are useful in large lecture theatres. The spheres are 50 mm or 64 mm in diameter and the rods are of plastic but otherwise the materials and method of assembly are the same as for the CCA kits, apart from the overall size of the model. A crystal systems kit (demonstration-size) and Bravais lattice kit (normal size) work on similar principles.

SASM also supply kits of parts for specific structures.

The Spiring Enterprises Ltd. Molymod assembled structures may also be purchased as kits containing the appropriate number and type of atom models and links. Quotations for other structures can be obtained on application. All those given in the current catalogue are (like the corresponding assembled models) very reasonably priced and the selection indicates that the Company is aware of recent trends in VIth form science teaching in this country.

## 4.4. NON-SPECIFIC COMPONENT KITS AND SEPARATE COMPONENTS

This section deals only with manufactured components of ball-and-spoke models and kits of components selected by manufacturers. Until recently, the contents of most kits of ball-and-spoke models seem to have been chosen chiefly to aid the teaching of elementary organic chemistry, though with the possibility of constructing one or two models of simple inorganic molecules. Many kits of space-filling, ball-and-spoke and skeletal models are still limited in application but several new kits have appeared in the last few years which are much more versatile. Perhaps the extensive range of assembled models provided by companies such as Crystal Structures Ltd. and Leybold has concealed a potential demand from inorganic chemists and crystallographers which is now beginning to be revealed in these days of financial stringency in educational institutions.

A list of suppliers of kits and components is given in Table 4.2. Suppliers' addresses may be found in Appendix 1. In the survey which follows no attempt is made to discuss every manufactured type of ball-and-spoke model. Rather, features of interest have been indicated and possible guidelines suggested to the would-be purchaser, who should have a clear idea of exactly what he requires of the models before making a decision, and then choose the best compromise.

Four important factors should be considered when selecting ball-and-spoke models. First the number and geometry of the drillings (or other points of attachment) in the spheres, secondly the nature of the ball/spoke link which

**Table 4.2**
Suppliers of ball-and-spoke models

| Name of model and/or supplier | Type available | Price code (Table 1.4) |
|---|---|---|
| Ace Plastic | c | a |
| Addatoms/SRM | n | a |
| Beevers | p, s, c | mostly a |
| Cenco-Peterson/Central Scientific | n | d |
| CSL | p, s, n, c | a, b, c, d,; mostly a, b |
| Disc/Addison-Wesley | n | a |
| Gallenkamp | n, c | a, b |
| Griffin and George | n, c | b |
| HGS/Maruzen/Addison-Wesley | n, c | a, b |
| Industrial and Scientific Instruments | n | a, b |
| Kristakit/Morris Laboratory Instruments | c | b |
| LaPine Scientific | n | a |
| Leybold-Heraeus | p | a, b |
| Macalaster Scientific | n | a |
| Molymod/Spiring Enterprises | p, s, n, c | nearly all a |
| PMDM/Prentice-Hall | n | d |
| Sargent-Welch | n | a |
| SASM | p, s, n, c | a, b, c, d |
| SRM | s, n, c | a, b |
| Unit/Cochrane/RJM Exports | c | a |
| Wells/Oxford University Press | n, c | a, b |

*Key*. Type available: p pre-assembled
s specific kit
n nonspecific kit
c components

determines the stability of the assembly and the ease with which it can be put together or taken apart, next whether free or restricted rotation about a link is possible and finally the scale. In the last case, visibility of a model in a large lecture theatre has to be set against the relatively high cost and storage problems.

### 4.4.1 Ace Plastic Company

A wide range of spheres in acrylic (Lucite, Plexiglas), cast phenolic resin (Catalin), cellulose acetate and other plastics can be obtained. Thirteen opaque and eight transparent colours are available in a range of sizes up to 1 inch and 0.75 inch diameter respectively. The spheres are not drilled but acrylic rod (1/8 to 1/4 inch) can be purchased and the spheres drilled by the customer. With larger spheres and other plastics the colours are limited. Bought in bulk the acrylic spheres are very cheap.

### 4.4.2 Addatoms

This inexpensive kit from Denmark is designed chiefly for the teaching of elementary organic chemistry. The atom models are pronged discs in five different colours. Two may be combined to represent another element. They are joined by fitting plastic tubes over the prongs. The number and arrangement of the prongs are suitable for the assembly of a reasonable range of organic molecular models. All the atom models are the same size and the tubes the same length. Two kits are marketed, the larger containing the same components in greater quantity.

### 4.4.3 Beevers

The 6.9 mm diameter spheres are made of methylmethacrylate and can be obtained in twelve opaque colours, to represent nonmetals, and eight transparent colours (metals). The standard drillings are each to 1 mm from the centre of the sphere and are octahedral (six at 90°), tetrahedral (four at 109.5°), trigonal (three in a plane at 120°), pyramidal (three at 109.5°) and one-hole. Other symmetries may be specially ordered. White opaque spheres with a diameter of 4.9 mm are used to represent hydrogen. The spheres are joined by stainless steel rods (1 mm diameter) which are inserted with the aid of pliers. Rods can be obtained ready cut in twelve different lengths representing common bond lengths on a scale of 10 mm ≡ 0.1 nm. These components are very moderately priced and the resulting models are of high quality.

### 4.4.4 Cenco-Petersen

Designed by Dr. Petersen [4] and marketed by Central Scientific Company, these models consist of neoprene spheres containing threaded aluminium inserts at the appropriate angles. The spheres are joined by stainless steel rods cut to length according to the scale used and bearing at each end, in effect, a locking nut device which is screwed into the aluminium insert of the sphere. By tightening or loosening this, restricted or free rotation about the link is possible. The spheres cannot come apart when joined in this way while strained structures can be built because the spheres are capable of distortion. Although strain is localised in a more rigid model, distortability of the spheres permits transmission of strain throughout the molecular model. A strain of up to 30° is possible and for this reason only relatively few atom models are required. For example, the trigonal carbon atom model can be used for ethenes, aromatic rings, carbonyls and carbonium ions. There are fifteen different atom models, three each of carbon and nitrogen, two each of oxygen and nitrogen, one octahedral metal, one hydrogen and one each of the halogens (excluding fluorine). Six lengths of precut rod are supplied with the kit. These correspond to five different carbon-carbon bonds and a single bond carbon-hydrogen for a scale of 50 mm ≡ 0.1 nm. Rods fifteen inches long are also provided which may be cut to correspond to other bond lengths. The univalent atom models are not flexible because they can only be used in terminal positions. Fifty-nine of these and fifty-five flexible atom models together with seventy precut rods are included in the set.

These models were further developed to include a reaction centre. This is discussed in Chapter 8.

### 4.4.5 Crystal Structures Ltd (CSL)

Precision models can be made with CSL wooden spheres and slit metal spokes. The spheres are painted according to the usual colour code and come in three sizes. 50 mm, 25 mm and 18.75 mm diameter. The depth of hole drilled in each is 12, 8 and 4 mm respectively. Holes in the large sphere take a 3/16 inch diameter rod while the others take 1/8 inch rod. Drillings are 1-hole, 2-hole (180°, linear), 3-hole (120°, trigonal planar), 4-hole (tetrahedral), 5-hole (trigonal bipyramidal), 6-hole (octahedral) and 8-hole (cubic). There are also two 12-hole drillings, one set along the six 2-fold axes of a cube and the other with hexagonal symmetry (six equatorial holes at 60° intervals, three in the upper hemisphere and three directly beneath in the lower) and lastly a 26-hole sphere with the drillings along all the symmetry axes of the cube. Most common symmetries except 3- or 6-fold can be achieved with the 26-hole sphere as well as some less common ones, such as the square antiprism.

Various kinds of spoke may be used to join the spheres. The simple, plain spoke is only suitable for permanent models because frequent removal and insertion of these would enlarge the holes in the spheres. Where repeated use is required a spoke with slit-ends which can be squeezed together for insertion but which expand inside the hole is used. This gives a tight fit without causing damage and is highly satisfactory, giving good service after many years' constant use. A variant on the slit spoke is the flexible spoke with a spring forming the central section. This can be used for multiple bonding (using two or three) or strained structures. Another with a larger pitch spring can be compressed or extended, so allowing the 'bond-length' to be varied. A spoke with a central swivel allows free rotation and one with a clamping device at either end prevents it.

Slit spokes in twenty-seven different lengths corresponding to the most frequently occurring bond lengths in organic compounds may be used to assemble molecular models of appropriate chemical species. The different spoke lengths are readily identified by means of coloured bands. The scale is 50 mm ≡ 0.1 nm. The Organic Chemistry Set (OCS) contains a selection of these spokes and suitably coloured spheres to go with them. Additional items such as five-membered rings, 'hedgehogs' or simulated van der Waals' radii are supplied separately.

A demonstration-size kit (LDS) employs 50 mm diameter spheres, 4 mm diameter spokes and a scale of 100 mm ≡ 0.1 nm, but otherwise the principles are the same as for the OCS.

The BS set consists of twenty-five 25 mm, 26-hole grey spheres and twenty-five similar spheres coloured red with seventy-five slit spokes 38 mm long and twenty-five of 63 mm lentgth. It is very useful for the assembly of a variety of binary crystal structures.

Other kits (SS1 and SS2) are designed for organic chemistry teaching in the VIth form. These are inexpensive and contain a useful number and range of components. SS1 has sixty-six spheres, a benzene ring and fifty-nine spokes while SS2 has twice as many components. There are also special kits for Open University courses.

Alternatives to wooden spheres and slit spokes are the self-coloured, polythene spheres (22 mm diameter) and flexible springs of the Linnell components. The spheres are in seven different colours (contrast this number with nineteen for the painted wooden ones) with 1-, 2-, 4-, 6- or 14 holes. The last is the most useful because it can be used for tetrahedral or octahedral arrangements and associated symmetries (but again not 3-fold). The Linnell Set (LNS) contains ninety-six spheres, forty 25 mm springs and forty 38 mm springs. It is cheaper than the other CSL sets being less precise and with simpler components but it is a useful general purpose set with sufficient similar components to build lattice models of a reasonable size. Extra components can be bought separately as with the other CSL models. The scale is of the order 30–40 mm ≡ 0.1 nm depending on which spokes are used.

### 4.4.6 Disc Molecular Models (Figure 4.5)

These are not dissimilar in appearance to the 'Addatoms' which they predate by about seven years. They are cheap but limited in scope and only relatively simple models of molecules or ions can be constructed with the kit components. The atom models are pronged discs, one inch in diameter, made from high impact polystyrene in twelve different colours. They can have one, two, four, five or six prongs. The angle between the prongs in a two-prong disc is approximately tetrahedral and the disc is used to represent divalent oxygen (red) or sulphur (yellow). For higher co-ordination numbers two discs have to be slotted together perpendicular to each other. This technique can give rise to tetrahedral, trigonal bipyramidal or octahedral arrangements of the prongs. The discs are joined by slipping tubing over the prongs. The tubing is mostly stiff but some flexible tubing is supplied so that two or three pieces can be used to represent double or triple bonds. Models of simple molecules can be assembled quickly and easily and the set could be useful to a 'beginning' student in chemistry for whom, presumably, it is intended.

Fig. 4.5 – Disc molecular models: (a) *trans*-dichlorobis(ethylenediamine) cobalt(III)cation (*trans*-$[Co(en)_2Cl_2]^+$), (b) methane.

### 4.4.7 Gallenkamp Linnell Models (Fig. 4.6)

These are exactly similar to the CSL Linnell models but additional types of component are available and there are several kits including a comprehensive demonstration set containing over five hundred each of spheres and springs. With this set a wide variety of molecular and crystal structure models can be made. In addition to the CSL components there is an 8-hole sphere for cubic co-ordination and spring lengths of 60 mm and 75 mm as well as 25 mm and 38 mm. The tapering at the end of each spring greatly facilitates insertion into the sphere and the resulting model is robust even when it is large. The components

Fig. 4.6 – Gallenkamp Linnell: (a) diamond ('d-glide' atoms in lighter colour), (b) rutile ($TiO_2$), (c) body-centred cube.

of the basic chemistry kit are almost identical to the CSL Linnell set except for minor colour differences and the fact that the Gallenkamp set contains two more 14-hole spheres. The 'crystallographic' set comprises one hundred-and-eight 14 -hole spheres in three colours and three hundred springs. Instruction leaflets are included with all the sets. All the components may be purchased separately. The sets were designed to meet the requirements of VIth form chemistry teaching in the UK but they are very versatile and can be used for many purposes. The overall approximate scale of a model using the two shorter spring lengths is the same as for the CSL Linnell models, 30–40 mm ≡ 0.1 nm. In molecular models the flexible springs may be used to demonstrate vibratory modes.

#### 4.4.8 Griffin and George (Fig. 4.7)

In my view the so-called 'Griffin Skeletal Models' are unambiguously ball-and-spoke models. However, it appears that the manufacturers use the term 'skeletal' simply to distinguish them from space-filling models.

The spheres are of hard, moulded, self-coloured plastics in two sizes. The larger ones (32 mm) are drilled in a variety of ways. Tetrahedral spheres come in black, blue, yellow or silver, trigonal bipyramidal in black or silver, octahedral in black, blue, yellow or silver, and both cubic and trigonal prismatic in either yellow or silver. Thus a useful range of co-ordination geometries is available. Extended arrays of the common binary ionic crystals can be constructed because tetrahedral, octahedral and cubic spheres are all available in more than one colour. The colour convention recommended by the manufacturers is that of the Colour Group of the Physical Society in which halogen is yellow and so there are no large green spheres. However, for MX crystals such as sodium chloride, caesium chloride, zinc blende and so on the co-ordination of each ion is the same. Here the colour is less important than the co-ordination geometry, providing more than one colour with the same drilling is available, as is the case with these models. Almost all the small spheres (19 mm, in nine different colours) have only one hole, though the red sphere has two with the tetrahedral angle between them. The spokes are of three kinds, flexible springs, rigid spokes (length 28 mm) or wooden rod which can be cut to size. The flexible springs are in two standard lengths, 48 mm and 83 mm, and these (in common with the rigid spokes) have specially designed ends which are similar in principle to the CSL slit spoke. These ensure a snug fit and minimum wear and tear on the drilled holes in the spheres. Although the manufacturers claim that this feature makes removal and insertion easy, I have found that removal sometimes necessitates the use of pliers. This is a good fault, leading to stability in extended arrays. Because flexible spring in 3 m coils can be cut to length and each end capped with a 'bond end' (available in packs of 50) if required, reasonably precise scaling is possible.

There are two kits. The simplest, intended for use in schools, contains twenty-four large spheres of assorted drilling geometry and colours, eighty-four small spheres and seventy-two springs 48 mm long. The 'Senior Set' comprises one hundred and four large spheres, seventy-eight small, forty-eight springs 48 mm long, twelve 83 mm long, thirty-six rigid spokes and some wooden connectors. Both include an instruction booklet but neither have spheres with trigonal prismatic drillings, long springs for cutting to size or 'bond ends'. These components have to be purchased separately.

The models are attractive, stable when placed on a shelf but rather 'floppy' when handled owing to the relatively large weight of the atom models and the flexibility of the springs.

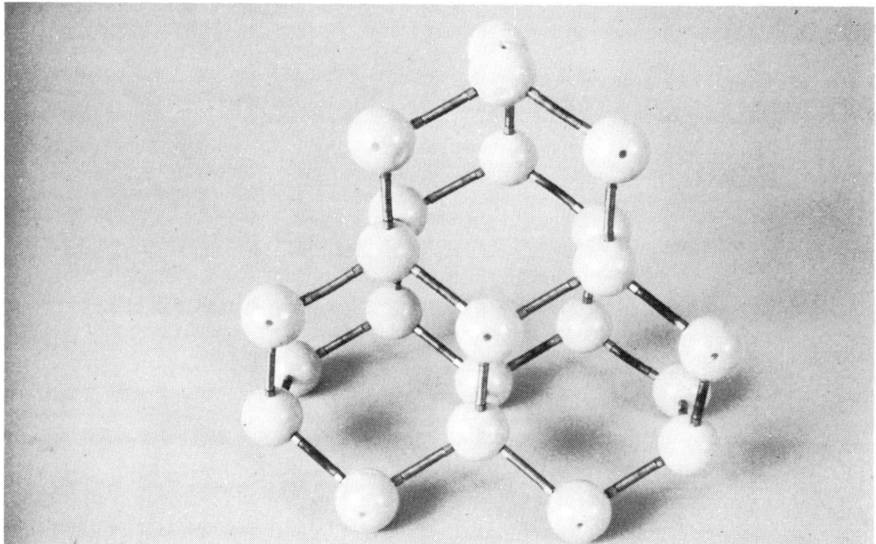

Fig. 4.7 – Griffin and George: zinc blende (ZnS).

### 4.4.9 HGS/Maruzen (Figure 4.8)

These Japanese models might very well be classified as 'polyhedra- and -plastic rod' except that only one other type (see PMDM below) would be in the same category. However their general appearance and their ethos are similar to those of ball-and-spoke models, the polyhedra simply being a device to facilitate location and identification of connector sites. Small spheres are used for hydrogen but polyhedra represent all other atoms. There are three kinds. The cubo-octahedron has fourteen faces and is derived from a cube by cutting off all the corners at the edge centres. The so-called '20-hedron' (not a regular icosahedron) is obtained by separating two halves of the 14-hedron by a planar hexagonal spacer. The 26-hedron is derived by truncating all the corners (twelve) of the 14-hedron. Effectively, the 14- and 26- hedra are analogues of the 14- and 26-hole ball as the drillings in the balls are perpendicular to the corresponding polyhedron faces. Therefore they have the same symmetry, the 26-hedron having faces perpendicular to all the symmetry axes of the cube. The 20-hedron is devised for structural arrangements with 3- or 6-fold symmetry. The atom models are of self-coloured plastics representing nine nonmetallic elements and one metallic element, the last in the larger of the two organic chemistry sets only. The inorganic set has carbon, hydrogen, and four metal or general purpose atom models while the crystal structure set has oxygen, sulphur, carbon and sixteen others of various colours and co-ordination, mainly tetrahedral and octahedral. These are described in the catalogue in terms of $sp^3$ and $d^2sp^3$ hybridisation although they are designed particularly for the construction of ionic crystal structure models. There are, in fact, over thirty polyhedral types, and a spherical hydrogen model, available as separate components. Because they have up to twenty-six holes it would be possible to construct most inorganic structures.

Connections are made by means of moderately flexible plastic rods, slightly tapered at the ends, which fit securely into the polyhedra and permit the assembly of strained structures. The models are quick and easy to construct and do not fall apart readily. Ten spoke lengths are provided for the organic chemistry sets proportional to the most frequently-occurring bond lengths on a scale of 25 mm ≡ 0.1 nm (common to all sets). For the inorganic and crystal structure sets another range of twenty-six lengths is supplied, covering not only a number of ionic bonds but van der Waals' distances and axes of unit cells as well. The longer rods (crystal structure set only) are of metal for greater rigidity. All components may be purchased separately.

The four kits range from a simple inexpensive set suitable as an aid in introductory organic chemistry teaching to one that is large and advanced, containing nearly nine hundred pieces. This is useful for constructing all kinds of crystal structure model and some macromolecules, because a hydrogen bonding model is manufactured.

Fig. 4.8 – HGS/Maruzen: (a) *(trans*-dichlorobis(ethylenediamine) cobalt (III)cation (*trans*-$[Co(en)_2Cl_2]^+$), (b) cyclohexane.

### 4.4.10 Industrial and Scientific Instrument Co., LaPine Scientific Co., Macalaster Scientific Co. and Sargent-Welch Scientific Co.

All these American companies offer one or more ball-and-spoke models kits. ISI's 'Student Molecular Model', LaPine's 'Organic Structure Set', Macalaster's 'Molecular Model Kit' and Sargent's 'Organic Structure Student Set' are all similar. They consist of fifty-four wooden balls painted with hard enamel paint in the appropriate colours, except for hydrogen, which is yellow. These represent seven elements, those for carbon and nitrogen having a diameter of 1-1/4 inch, the others 1-1/8 inch. Three types of spoke are included. There are wooden pegs 2-1/8 inch long, shorter ones of 1-1/4 inch and 2 inch metal springs for use in the construction of rings and multiply bonded species. The sets are mostly useful for the study of elementary organic chemistry but models of some simple inorganic molecules can be constructed.

A somewhat larger set can be obtained from ISI and LaPine which contains one hundred balls and eighty-five springs in three lengths, 1 inch, 1-1/8 inch and 1-1/2 inch. Six types of atom model are included, the one intended to represent nitrogen being painted orange. Again, the emphasis is on organic chemistry at a fairly elementary level.

Two further kits are available from ISI, both large scale demonstration size, one for general chemistry and the other for organic. The former has a fluorine and four metal atom models, as opposed to one metal model in the organic set. Otherwise the elements represented are the same, though not the quantities of each kind of atom model. Both sets contain three 'orbital lobes' which may be used to represent lone pairs of electrons. Sticks are inserted into previously installed vinyl grip fasteners to join the atom models, and the organic chemistry set has extra unassigned spheres together with 'bond angle locators' so that they can be adapted for any specialised application.

The sets are not dissimilar to the earlier Science Related Materials (SRM) kits (see below) which have now been superseded.

### 4.4.11 Molymod (Figure 4.9)

These inexpensive models are sold assembled (Section 4.2.6 above), in specific kits (Section 3 above), in nonspecific kits and as components.

There are six sets, 'Standard' or 'Introductory', 'Senior' or 'Organic', 'Inorganic/Organic', 'Space-filling', 'Complex-Ion' and 'Biochemistry'. In addition there is a game called 'Atoms' which uses Molymod components. The number of atom models and links in each set respectively are as follows; atoms/links 48/62, 97/140, 93/86, 136/128, 151/158, 252/200 and 44/50. There are thirty-two different atom models made of hard, self-coloured polystyrene. These are slightly truncated perpendicular to the holes provided for link insertion but are generally approximately spherical in appearance. The usual colour code is used. There are short, medium and long polythene links which fit reasonably tightly into the holes in the atom models. They are flexible and, like springs, can be used to indicate multiple bonding, by using two or three to join a pair of atom models, or to construct strained structures. All the kits except the 'space-filling' one contain an instruction leaflet or booklet.

The Molymod system is geared to the requirements of VIth form chemistry teaching in the UK, the 'Complex-Ion Set' being specially designed for the Nuffield syllabus. Its use outside this field is somewhat limited.

Fig. 4.9 – Molymod showing medium and short links: (a) *meso*-2,3-dibromobutane, (b) *n*-pentane, (c) *iso*-pentane, (d) *o*-nitrophenol.

**4.4.12 Polyhedral Molecular Demonstration Models Set (PMDM),** (Figure 4.10).

When this set first appeared a few years ago, a number of educational institutions might have been able to afford it. However, now it would be too expensive for most. This is a pity as it is an attractive and useful kit with many admirable features. While really large models can be built using the full length (15 inch) of the longest connector tubing supplied, they can be scaled down as required by cutting shorter lengths or using the standard 3.5 inch lengths included in the set. Of the hundred short connectors, seventy are rigid and thirty flexible while sixty long connectors are rigid and forty flexible. The flexible connectors are used for strained structures and multiple bonding, as with many other ball-and-spoke models, but also for outlining van der Waals' envelopes and electron probability distributions in molecules with lone pairs or $\pi$-bonds. Coming in several colours flexible tubes can either correspond to the atom to which they are attached or can indicate 'orbital lobes' of opposite sign. The flexible tubes are attached by means of T, Y, X or 'universal' pieces, the T-shape being used for van der Waals' contours because the result is an approximately spherical outline. The 'universal' piece consists of six equally spaced arms in a plane and has several applications. The rigid tubes are pressed manually into metal lined connector sockets which are a little too small for them. In this way the tubing is crimped at the end and tightly held. In fact, I experienced some difficulty in inserting and removing the tubes. Of course, it is particularly important that the assembly is strongly held together in a demonstration-size model and this is certainly the case with PMDM models.

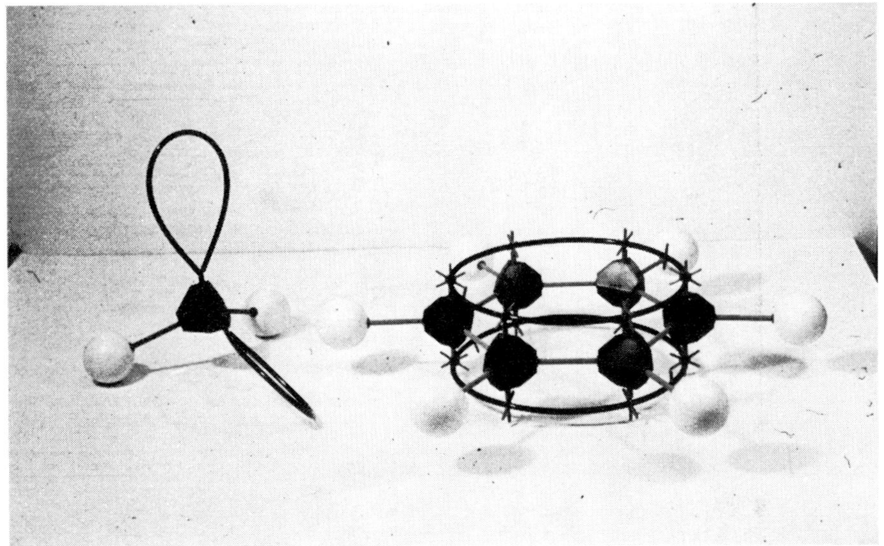

Fig. 4.10 – PMDM: (a) water, (b) benzene.

The atom units are either spheres or polyhedra. There are fifty of the former each with one hole, representing univalent atoms, and colour coded in eight colours. The fifty-six polyhedra comprise thirty-five tetrahedra, seven octahedra and fourteen trigonal bipyramids, each supplied in black, blue, red, yellow and grey. All the atom units are hollow and of light, self-coloured plastics.

The components are conveniently packaged in a two-tier polystyrene box and an instruction manual is included.

### 4.4.13 SASM

The SASM 'Modèles Moléculaires Universels' are well-named because the number of possible permutations seem to be almost infinite! As well as the assembled models and specific kits already described, many kinds of kit or separate component can be bought. Sets are available for teaching or research with several different scales. Atom models can be spherical or truncated and of plastics or wood. The spokes can be of metal or plastics, flexible, curved or dumb-bell shaped and permitting free or restricted rotation.

### 4.4.14 Science Related Materials (SRM)

Assembled models and 'Crystal Construction Atoms Sets' (specific kits), normal and demonstration sizes, are discussed above in the appropriate Sections, but SRM also markets nonspecific ball-and-spoke kits in normal and demonstration sizes. 'Molecular Construction Atoms' come in a 'Student Set' or a 'Research and Teaching Set', both with a scale of 30 mm ≡ 0.1 nm. There are sixteen different types of colour-coded expanded polystyrene sphere with the bond positions marked ready for piercing with the special tool provided. Apart from hydrogen (yellow) and silicon (yellow-green), the conventional colours are used to represent the elements. However, where more than one atom type is used to represent one element, a related colour is used. For example, carbon is black or grey and nitrogen is blue or blue/green. There are three different sphere diameters, 13 mm (hydrogen only), 35 mm (sulphur, phosphorus, chlorine, silicon and metal) and 25 mm for all the others. Once the holes in the spheres have been made, vinyl sleeves are cemented into them. The spheres are then joined by colour-coded, wooden rods, precut to scale. For permanent models, the rods may be glued into the spheres without lining the sockets with vinyl. Although the sockets are arranged so as to give the common symmetries, the kit is better suited to the construction of models of organic rather than inorganic species, in common with many others. The inexpensive student set contains one hundred spheres while the research and teaching set contains one thousand. The smaller is a useful general purpose set for beginners.

The spheres for the two demonstration-size 'General' and 'Organic Chemistry' sets range in size from 51 mm to 81 mm diameter and the lengths of the connecting rods from 25 mm to 230 mm. Ten elements are represented by sixteen types of sphere, including three metal atom models. All are appropriately colour-coded and of expanded polystyrene. 'Vinyl receptors', holes lined with vinyl for the reception of a connecting rod, are installed by the manufacturers in the appropriate positions in the spheres. All the user has to do is push the rod into the hole, where it is held strongly though allowing a little flexibility in the assembly. The plastic connecting rods are also colour-coded to help in the quick identification of single, double and triple bond lengths while assembling a model before a class. Prebent bonds (rigid) and flexible bonds permit the construction of strained and multiply bonded species, the latter being indicated by more than one bent bond or by a single shortened one as required. The models are intended for use with chemistry courses for nonspecialists at first year university level and are excellent for this purpose although they are rather expensive.

### 4.4.15 Unit Models (Figure 4.11)

These are relatively new ball-and-spoke models from RJM Exports Ltd. who until recently have specialised in several kinds of skeletal model (Chapter 5). There are three components, spheres, pegs and tubes. The large (45 mm diameter) spheres are hollow, of hard plastic (polythene), and contain enough holes to give the common symmetries and co-ordination numbers from one to twelve. The sphere comes in two halves which can be locked together in three orientations, 0°, 45° and 60°. At 0° the two sets of holes are aligned. At 45° there are holes in a square antiprism arrangement, while at 60° a twelve co-ordinate arrangement such as occurs in cubic close-packed structures is possible. Short pegs are inserted into the holes to give the required symmetries and the spheres may then be joined by sliding connector tubes over the pegs. The spheres are available in seven colours. A smaller sphere (25 mm diameter) has one hole and is only produced in white. It is used to represent hydrogen.

Fig. 4.11 – Unit: (a) carbon tetrachloride, (b) sulphuric acid.

There are four types of peg. One is used for radial positions, a second for polar positions, the third for tetrahedral or 8-co-ordinate species, and the last for 12-co-ordination. The polar peg is linear but the others are angled so that although all the holes moulded into the large sphere are parallel, the pegs project along radial lines as is necessary. Tubes may be rigid or flexible, the latter being for multiple bonds. Unfortunately the spheres are too heavy for the 'rigid' tubes in many assemblies, their weight sometimes causing the tubes to bend, kink or split. The suppliers recommend that a wooden dowel should be inserted into the tube. Also, the part of the peg which projects from the sphere is rather short and the tube occasionally slips off. The recommended scale is 60 mm ≡ 0.1 nm, which is demonstration-size, but the tubes can be shorter of course.

A set is now available which comprises fifty large spheres in assorted colours, two hundred assorted pegs, one hundred and forty 50 mm tubes and twenty-eight 210 mm tubes. In addition, packs of spheres either of a single colour or of several colours, pegs of one or several types and bundles of tubes may be purchased separately.

#### 4.4.16 Wells Models (Figure 4.12)

Marketed by Oxford University Press, the Wells kit contains several model types including spheres for close-packing, skeletal, polyhedra and ball-and-spoke models. The last category consists of one hundred 2 inch spokes (plastics rod) and forty spheres, twenty red and twenty grey, all with the 26-hole drilling, along the symmetry axes of the cube. They are used to construct models of various inorganic crystal structures and to illustrate structural principles. The Wells kit is considered in more detail in Chapter 6.

Fig. 4.12 – Wells: $Si_6O_{18}^{12-}$ anion in beryl($Be_3Al_2Si_6O_{18}$).

### 4.5 HOMEMADE AND MISCELLANEOUS BALL-AND-SPOKE MODELS

#### 4.5.1 Jigs, Templates and Punches

The most difficult feat in the preparation of homemade ball-and-spoke models is location of connector sites on the spheres so as to achieve a reasonable approximation to the correct symmetry. Sanderson [5] believes that 'a good eye for symmetry' can make any construction aid unnecessary but considerable experience is essential if such expertise is to be acquired. The Griffin paper protractor template sold with the Grifzote spheres (Chapter 2 and Table 3.1) is one of many similar devices which can help in 'bond' location. References [5] to [13] inclusive at the end of this Chapter all provide detailed information for the preparation and use of simple jigs or templates which can be made from cheap commonly available material with minimal skill and equipment.

Typical of many is that described by Bassow [6] for the location of up to four sites tetrahedrally distributed. The jig is made from thick packing-case (corrugated) cardboard. A circle with the same diameter as the sphere to be used is marked on the cardboard and two lines are drawn from the centre of the circle with the required angle (for example 105° for a water molecule) between them. The circle is then cut out and the sphere is inserted into the hole so that the edge is level with its equator (mould marks sometimes indicate where this is). Wire punches are made from 1/16 inch welding rod, for example, bent round at one end so as to form a handle and sharpened at the other. The length is such that when the punch is pushed through the cardboard layer up to the handle, the sharp end will penetrate to the centre of the sphere. For a regular tetrahedral molecule such as methane, the angle between the guidelines would have to be adjusted to 109.5°. Having punched two holes in the sphere, this would then have to be rotated through 90° about its vertical axis and two further holes punched. Sharpened wooden dowels can be inserted into the holes made by the punches and the model completed by the addition of four spheres representing hydrogen. Variants on this type of construction aid are not described here because most readers will have access to one or more of the references cited, but more information on construction devices generally is given in Chapter 9.

However, there are some rather more sophisticated ball punches such as the 'Kristakit' designed by Ladd [14] which is commercially available (Figure 9.1). Others have been made in school and university workshops though the designs have not all been published. These are sometimes of metal, easier to use and more precise than the simple models. As the 'Kristakit' was designed specifically to be used in the construction of ball-and-spoke lattice models of inorganic crystal structures it is described here. Only one size of sphere is necessary for lattice models and one inch diameter expanded polystyrene spheres are used. Although any arrangement of punched holes can be achieved, detailed instructions are given for tetrahedral, octahedral and cubic arrangements because the emphasis is on the construction of three-dimensional arrays. Tightly held in position by a spring-loaded clamp, the ball can be rotated together with a protractor (an integral part of the apparatus) to the required angle relative to the push-rod which makes an indentation at the appropriate point on the sphere. The spheres are joined by 1/8 inch diameter wooden dowel rod which can be cut to size with the guillotine provided, the recommended scale being 25 mm ≡ 0.1 nm. The guillotine comprises a linear scale parallel to two grooves into which the dowel rod fits, a slide to mark off the required length and a blade on a swivel which can be swung sharply on to the rod so severing it neatly.

In addition to the items mentioned, one hundred and fifty balls and two hundred feet of rod, the kit contains a very helpful instruction booklet which gives clear and detailed directions for the use of the ball-punch and the assembly of several models. These include α-iron, caesium chloride, sodium chloride, diamond, zinc bende and several molecules. A relatively advanced exercise demonstrates how a model of rutile can be constructed from a knowledge of the crystallographic data. This is preceded by sections on the determination of bond lengths, angles, location of atoms in unit cells and stereographic projections. Other data are given in appendices. The 'Kristakit' can be used by teachers or students, the degree of involvement of the latter depending on the extent of their knowledge. While spheres punched by the teachers could be assembled by students at a very elementary level, construction of the rutile model would present quite a challenge to university students versed in structural inorganic chemistry.

### 4.5.2 Adaptation and Invention

The models enthusiast should always be a snapper-up of unconsidered trifles [15]. It is amazing how many items discarded by other people can be put to good use. One usually fruitful source is the children's toy-box. Many toys manufacturers produce all kinds of variant on the ball-and-spoke theme. One such is 'Fixaball', from which I constructed a model of perovskite, and there are numerous others. Buseck [16] used 'Tinker Toys', which consist of rods and discs, to construct models of crystal lattices for teaching symmetry properties in his introductory mineralogy course. Collard and his co-workers [17] devised some models to be used in a programmed learning course in organic chemistry. Their spheres were formed from two interlocking hemispherical polystyrene shells of the type used in babies' rattles. Several sizes and colours could be obtained. The hemispheres were drilled and then the two halves glued together. Plastic sockets were glued into the holes and the spheres joined by rigid or flexible plastics rods with slit and barbed ends which locked into the sockets. The students were issued with sets and assigned model building tasks as part of their learning programme.

Some models made from spiral spring curtain 'rods' and poppet beads have been described as remarkably economical [18] while another example of adaptation of domestic materials is the use of carpet press-studs by Turner [19]. The aluminium connecting rods were drilled at both ends and the male ends of the press-studs inserted. The female ends of the studs were fitted into polystyrene spheres in the appropriate positions. When the stud was joined, free rotation about the spoke could occur.

The Griffin Chemical Structures Frame (Figure 4.13) is a manufactured article consisting of a large framework cube incorporating eight smaller cubes with sides half the length of the large one. Each has a body diagonal, to which can be attached balls of plasticine in atomic (or ionic) positions so that the overall effect is of a ball-and-spoke assembly. Urban beach-combing on my part produced a lamp-shade frame, similar to the Griffin framework, which has since been converted into a model of a 4-fold screw axis.

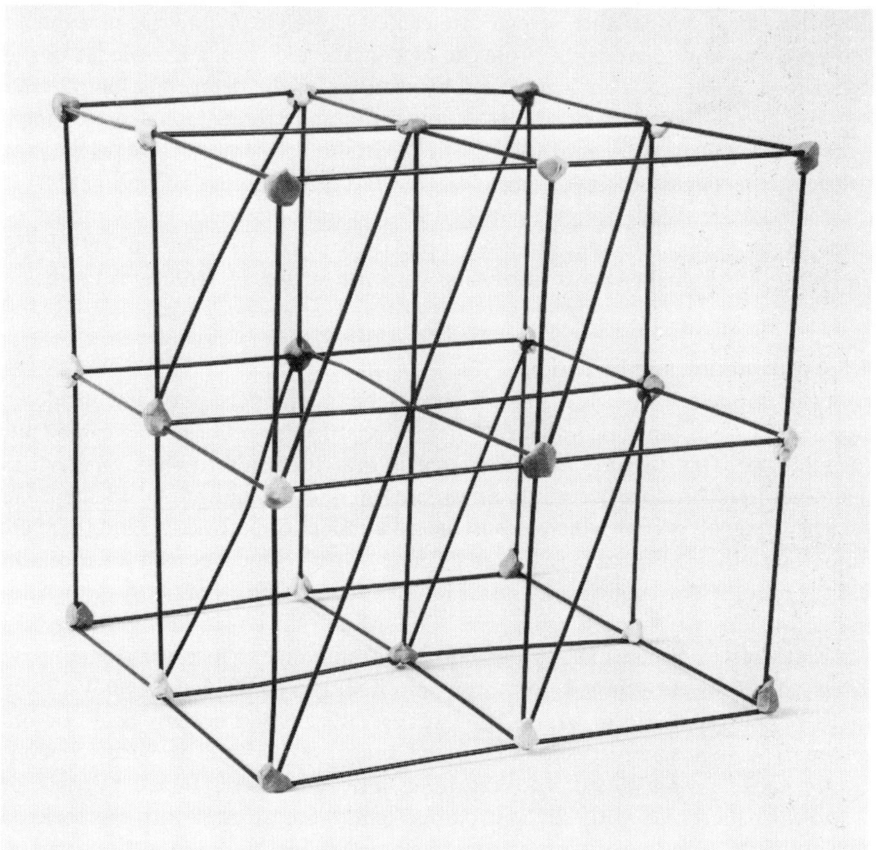

Fig. 4.13 – Griffin and George: Chemical Structures Frame, sodium chloride lattice.

Similar in principle to the Chemical Structures Frame are the models designed by Craig [20] to demonstrate molecular symmetry. These consist of spheres mounted, by means of brass rod, in holes in transparent plastic sheets representing planes of molecular symmetry. The spheres indicate atomic positions. Similar to these is Olsen's model of a face-centred cube [21] in which cork balls are cemented to 'plexiglas' sheets. A 'plexiglas' cube can also be made and partial spheres cemented on at the appropriate places. For example, half spheres can be placed at face entres and one eighth of a sphere at each corner.

Subluskey [22] has designed ball-and-spoke models in which precise variation in bond angle is made possible by a 'bond restraining clip'. The spheres are of rubber with the holes moulded in while the spokes are of aluminium enlarged at each end so as to ensure a tight fit. In the water molecule, for example, the bond angle is 105° but a sphere with its holes tetrahedrally disposed can still be used. The two spheres of the molecular model are drawn towards each other by means of the bond restraining clip which clamps on to each, reducing the angle from its 'ideal' tetrahedral value to a known fixed value (in this case 105°) determined by the dimensions of the clip. Subluskey also used two bent spokes for multiple bonds, an early example of this currently common practice.

## REFERENCES

[1] Muetterties, E. L. and Wright, C. M., *Q. Rev. Chem. Soc.*, **21**, No. 1, 109, (1967).
[2] Institute of Physics, *J. Scient. Instrum.*, **24**, 249, (1947).
[3] Beevers, C. A., *Educ. in Chem.*, **11**, 198, (1974).
[4] Petersen, Q. R., *J. Chem. Educ.*, **47**, 24, (1970).
[5] Sanderson, R. T., *Teaching Chemistry with Models,* Van Nostrand (1962).
[6] Bassow, H., *Construction and Use of Atomic and Molecular Models,* Pergamon (1968).
[7] Bentley, C. S., *School Sci. Rev.*, **48**, 488 (1967).
[8] Conard, C. R. and Bent, H. E., *J. Chem. Educ.*, **46**, 492, (1969).
[9] Dabby, R. E., *Making Crystal Models,* Pergamon (1969).
[10] Greenwood, N. N. and Smith, J., *Educ. in Chem.*, **1**, 25, (1964).
[11] Nuffield Chemistry: *Handbook for Teachers,* Longmans/Penguin (1967).
[12] Spargo, P. E., *J. Chem Educ.*, **51**, 259 (1974).
[13] Tetlow, K. S., *Educ. in Chem.*, **1**, 7, (1964).
[14] Ladd, M. F. C., *Educ. in Chem.*, **5**, 186, (1968).
[15] Shakespeare, W., *The Winter's Tale,* Act IV, Scene 2.
[16] Buseck, P. R., *J. Geol. Educ.*, **18**, 26, (1970).
[17] Collard, M., Griffith, J., Liddy, H., Shuk, V. and Swinbourne, E. S., *Educ in Chem.*, **6**, 130, (1969).
[18] Stewart, J. R. and Sherwood, M., *Educ. in Chem.*, **4**, 286, (1967).
[19] Turner, M., *J. Chem. Educ.*, **48**, 407, (1971).
[20] Craig, N. C., *J. Chem. Educ.*, **46**, 23, (1969).
[21] Olsen, R. C., *J. Chem. Educ.*, **44**, 728, (1967).
[22] Subluskey, L. A., *J. Chem. Educ.*, **35**, 26, (1957).

# 5

# Skeletal models

## 5.1 INTRODUCTION

Moving from space-filling models, through ball-and-spoke to skeletal, there is a shift in emphasis from the atoms or ions to the bonds joining them. In a typical skeletal or framework model an atomic or ionic position is only indicated by a point at which two pieces of the framework (usually rod or tube) are joined or intersect. Consequently these are the most 'open' of all models and measurements of dimensions and angles (including dihedral angles) are greatly facilitated by the accessibility of any part of the model to measuring instruments. Inevitably steric hindrance is less obvious than in any other type of model, but several methods exist for cladding the bare framework with various materials or for attaching devices which indicate the extent of the van der Waals' envelope. Great diversity is manifested by skeletal models. On the one hand, extremely precise models are used extensively for research purposes, while on the other the simplest are used in a variety of teaching applications.

The most precise are employed chiefly in the portrayal of covalent structures and in particular large organic molecules of biochemical importance. In this context they are valuable in studies of reaction mechanisms and conformational analyses where, because of their accuracy, subtle structural features are revealed. On some occasions precise skeletal models can enable the investigator to predict or confirm molecular structures. Examples are given in the appropriate sections of this Chapter. In my experience, such models are the type most frequently mentioned in the chemical research journals, and always in an organic chemistry application because the precise models are specialised and necessarily less versatile than the simple ones. In many respects they are complementary to the most precise space-filling models, often being designed to represent similar structures. Therefore one would ideally like to be able to compare a space-filling and skeletal model of the same molecule.

Although all skeletal models have in common the ability to highlight the geometrical pattern of bonds in a molecule or crystal, within this category great variation in precision, complexity, construction principle, technique of assembly and range of application is possible. Metal rods or tubes accurately proportional to internuclear distances may be replaced by wooden or plastics rods or tubes which may or may not be accurately scaled. In many cases, plastics tube or rod measured with a ruler and cut with scissors can be assembled to give perfectly adequate models, particularly where all the bonds are the same length, as in ionic structures. With these simpler model building systems, what is lost in precision is gained in versatility. The earlier manufactured models of this type were designed primarily to represent organic structures but they could be adapted to some extent so that their range of application was widened. However, in recent years manufacturers have apparently become aware of a demand for models which could be used for the portrayal of inorganic compounds, including ionic crystal structures. As a result several new types have appeared capable of meeting these requirements, presumably generated by changes in emphasis and course content in chemistry teaching at secondary school level. At the tertiary level, I have made extensive use of skeletal models as discrete polyhedra of varying complexity and to demonstrate linkage of simple polyhedra in chains, sheets or three-dimensional arrays. Discrete polyhedra can be used to show symmetry properties, chirality and less common co-ordination numbers. They can also demonstrate relationships between 'idealised' geometries such as the monocapped octahedron and pentagonal bipyramid, where a moderate distortion of one arrangement can make the other appear to be a more appropriate description, albeit still distorted. While the structural chemistry associated with linked tetrahedra and octahedra is discussed in Chapter 6, linkage of less symmetrical polyhedra can be conveniently demonstrated by use of certain types of skeletal model. For example, it can be shown that the trigonal dodecahedron can form chains but cannot form a three-dimensional structure. The 'openness' of the framework model is valuable in discussions of those ionic crystal structures in which one or more of the ions may be regarded as being close-packed. For instance, using these models it is easy to see that a hexagonal close-packed framework consists of two interpenetrating hexagonal prisms, one representing the 'A' layers and the others the 'B' (Figure 5.1).

Ways of representing giant molecules, polyions and extended arrays are considered in Chapter 6 and some of these employ skeletal models. Therefore these types of model are described here but their application is discussed in the next Chapter. This is because, although there is clearly a need among biologists, biochemists and environmental scientists for information concerning models of macromolecules, they are not necessarily interested in other kinds. I hope that a gathering together of the various designs and techniques for the portrayal of these large chemical species will prove to be helpful. Cross references are given so that as little material as possible is duplicated.

Distinctions between skeletal and other model types are again arbitrary and inevitably to some extent subjective.

Fig. 5.1 – Orbit model of interpenetrating hexagonal prisms.

## 5.2 COMMERCIALLY AVAILABLE SKELETAL MODELS

### 5.2.1 Atomunits (Figure 5.2).

These have been classified as skeletal because the manufacturers (Capital Biotechnic Developments Ltd.) describe them as such but a case could be made for placing them in the ball-and-spoke category. The coppered mild steel 'bond rods' which form the framework are joined by moulded, self-coloured plastics 'atomunits'. These units consist of a cluster of seven tubes, four of which are circular in cross-section and disposed tetrahedrally, while three have square cross-sections and point towards the corners of an equilateral triangle. In consequence the following angles are available, tetrahedral 109° between round tubes, trigonal 120° between square tubes and 180°, 94° and 105°, each between one square and one round tube. These enable models of most organic compounds to be constructed, especially those of biochemical interest. For example, the 94° angle is found in sugars and 105° in the water molecule. The lack of an octahedral arrangement limits the usefulness of the 'atomunits' in inorganic chemistry. The units are produced in six colours, black (C), red (O), blue (N), yellow (S), purple (P) and white (unspecified). A round white marker is used for terminal hydrogen and a white-sleeved rod for a hydrogen bond, while terminal halogens are indicated by light or dark green round markers. All these represent atomic centres.

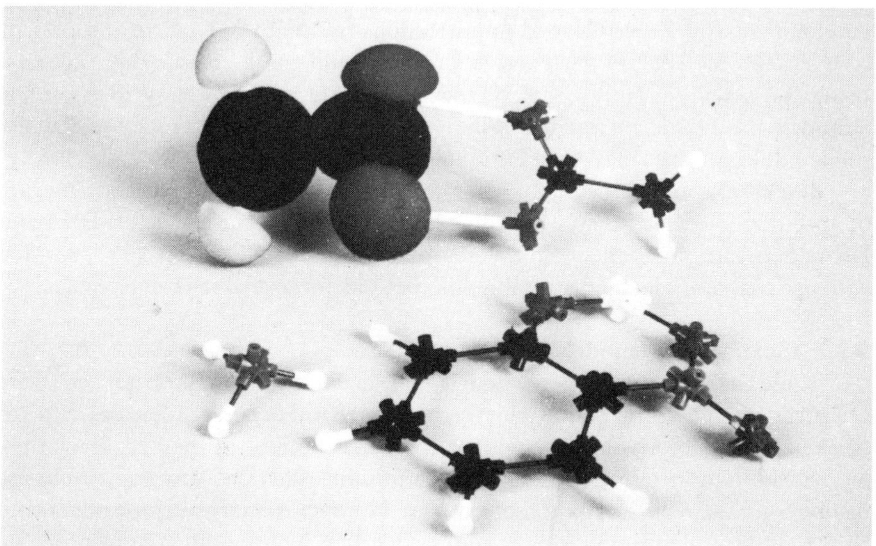

Fig. 5.2 – Atomunits: (a) acetic acid dimer with cladding 'mushrooms', (b) ammonia, (c) *o*-nitrophenol.

The 'atomunits' are joined by the 'bond rods' precut in six different lengths proportional to the most commonly occurring bond lengths in biochemical structures, except that allowance is made for the depth of penetration of the rod into the 'atomunit'. The rod stops 2 mm short of the centre. In addition there are two types of bent rod which may be used for multiple bonds, two of one type for double bonds, three of the other for triple bonds. This provides a hint of the electron density distribution, but an alternative technique is to use a straight bond of appropriate length with a square sleeve 'bond lock' placed over the connecting rod and square section 'atomunit' tubes at either end to prevent rotation about the 'bond'. The scale obtained with the precut rod is 20 mm ≡ 0.1 nm but uncut rod may be purchased together with special pliers incorporating a gauge so that larger scales are possible. Two sizes of support frame are manufactured to which large or small 'atomunit' assemblies may be attached for display purposes.

A useful feature of the 'atomunits' system is that cladding to give a space-filling effect may easily be accomplished by the attachment of foam plastics colour-coded mushroom shapes to vacant arms on the 'atomunits'. If one of these is attached to every atomic centre and all are placed on the same side of the molecular model, a fair indication of the shape of the van der Waals' envelope is given and an approximate overall scale of 20 mm ≡ 0.1 nm is maintained. It is frequently convenient to 'space-fill' in this way just that part of the model where steric hindrance might be particularly significant.

Homemade precursors of these models were designed by Doré [1]. Further developments by Doré [2] and modifications by Smith [3] led to the present system. The models are inexpensive and light, and can be reasonably accurate. Using the simple assembly tool provided they are relatively easy to assemble, although the plastic material is rather inflexible. Models of moderate size are stable but larger ones may need extra mechanical support.

Several kits are marketed both for general purposes and for the construction of specific molecular models including various proteins, nucleic acids, DNA and RNA. Separate components may also be purchased. A booklet [3] provided with each kit gives detailed assembly instructions for several macromolecules.

### 5.2.2 **Dreiding Stereomodels** [4] (Figure 5.3)

These beautiful precision models are made of stainless steel and were designed for the construction chiefly of models of organic molecules and, in particular, large biochemically significant species. Because of their accuracy they are widely used in research, their most important functions being to emphasise distinctions between similar structures, for example isomers or conformers, and to suggest possible reaction routes and mechanisms by use of models of probable transition states.

Quantitatively, the Dreiding models are an aid in conformational analysis where energy contents and dipole moments of different conformers have been calculated. Robinson and Theobald [5] used these techniques in their studies of cyclohexane derivatives. The dipole moments of all plausible conformations of a particular molecule were calculated from the Dreiding models and that which corresponded to the experimental value was assumed to be correct. In this case the conformation which was shown by the model to accommodate the calculated dihedral angles most readily was correct. Dipole moments calculated from measurements on Dreiding models were also used by Collins and Kirk [6] to determine the magnitude of polar effects in some cyclohexane derivatives. Atomic co-ordinates have been determined and dipole moments calculated using a technique described by McEachern and Lehmann [7] in which shadow projections of a Dreiding model were obtained. A combination of Dreiding and Framework Molecular Models (see 5.2.4 below) was used by Chipman [8] for the construction of strained structures. The models were then used to illustrate the Woodward-Hoffman rules for the conservation of orbital symmetry. Dreiding models are also useful in studies of chiroptical properties and the interpretation of nuclear magnetic resonance spectra.

Fig. 5.3 – Dreiding: testosterone.

The Dreiding units consist of one or more stainless steel rods or tubes brazed together with the appropriate angles between them. A single aliphatic carbon unit is formed from two rods and two tubes tetrahedrally disposed. The centre of the tetrahedron is painted black because this indicates the position of the carbon nucleus. The distance from this point to the end of a rod or a tube, representing the hydrogen nucleus, is proportional to the C-H bond length. Therefore the unit actually represents a molecule of methane. When two such units are brought together, a tube is inserted into a rod until a projection on the inside of the tube snaps into an indentation on the rod. The distance between the two unit centres is then proportional to the single C–C bond. Joining a C-H and an H-O gives a C-O single bond and similarly a C-H and an N-H give a C-N. The manufacturers (Büchi) call this process ‘dehydrosynthesis’ because in each case two lengths, each corresponding to the covalent radius of hydrogen, are eliminated and the length left is equivalent to the sum of the covalent radii of the bonded atoms.

While the linkage mechanism firmly holds the units apart at the correct distance, free rotation can occur about the link providing it is linear and unstrained. The scale is 25 mm ≡ 0.1 nm, which means that even models of macromolecules are not too large and the assembly is stable. Each pair of multiply bonded atoms is represented by a single unit so that no rotation about the bond is possible. An aromatic ring is built from three such pairs. Sleeve coding is used to distinguish different types of double bond, an isolated carbon/carbon double bond having one marker ring, a conjugated double bond two, an aromatic none and so on. Without these markers mistakes could be made because small differences in lengths and angles cannot always be readily discerned. Points of intersection of rods and tubes indicate positions of atomic nuclei and these points are painted different colours for easy identification. The usual colour code is used for the most frequently represented elements, except that for the halogens beige, yellowish-green, reddish-brown and dark brown indicate fluorine, chlorine, bromine and iodine respectively. No colour is applied for hydrogen but the metals are variously indicated by greens, greys, blue and yellow.

Approximately one hundred and fifty different units are manufactured in order to provide precise components for a wide range of structures. Although the univalent species have to be supplied in both rod and tube form, which accounts for about thirty units, there are still many more different pieces than are available for any other kind of molecular model. The Dreiding models cater particularly well for cyclic systems, three- to seven-membered rings being provided for as well as heterocyclics including fused rings with both endo- and exo-conformations. Saturated and unsaturated ring species can be constructed. Special units are employed where strain occurs, where the angle is 'squeezed' in three-, four- or five-membered rings and where the angle is 'spread' in seven-membered rings and some chains. The three-ring may be purchased as a single unit. In the somewhat unlikely event that a suitable unit for a cyclic system is not available it can be specially prepared by the manufacturers. In fact the Dreiding models can withstand a few degrees of strain without permanent deformation, but greater distortion may cause them to be irreversibly damaged. Because units representing six-co-ordinate magnesium and the first transition series from manganese to zinc are manufactured together with four-co-ordinate tin, lead, zinc (tetrahedral) and nickel (square planar), a number of organometallic systems can be built. A complete ferrocene model is featured in the catalogue and assembled models of macromolecules such as polypeptides, DNA and vitamin $B_{12}$ can be obtained on request.

Kits for building a range of model types can be supplemented by those for three-, four-, five and seven-membered rings. In addition peptide, nucleic acid and porphyrin kits may be purchased. These are good but expensive.

There are various accessories. The Angstroemscala is hinged at its zero point for attachment to a tube or rod so that distances to other parts of the molecular model can be read in Angstroms and angles can be measured with a torsion angle measuring device. It is perhaps appropriate at this point to mention devices designed by ApSimon and his co-workers [9] for measuring angles (including dihedral angles) and distances in Dreiding models. The papers on this topic describe a protractor arrangement about which an extendable pointer pivots. One half of an interlocking unit clamps on to the middle of a 'bond' in a Dreiding model while the other half, bearing the pointer, can rotate about the longitudinal axis of the 'bond' and the angle of rotation can be measured. The pointer consists of a sliding rod within a tube the end of which can be moved to the distant group. The rod is then clamped and the length from tip to pivot measured. The designers were investigating the effect of anisotropic groups on long range shielding in the nmr spectra of diterpenes and steroids and needed accurate estimates of intramolecular distances. In this way the distance of the 9-methyl group from the carbon-oxygen double bond was measured in 9-methyl-$\Delta^2$-octalin. A development of this device can be attached by a specially designed pivot at the centre of a benzene ring.

The other Dreiding accessories are, first, friction springs which can be placed over 'bonds' reducing the rotation facility where it is necessary to hold the model in a particular conformation. A reconditioning tool is available which renews indentations in the snap-lock mechanism when they have become worn with use and are no longer reliable. 'Colormark' spheres can be attached to terminal groups where these need to be easily identified, for example when the conformation of the model is being changed or when a transition state is being examined. Finally there are also van der Waals' hydrogen radii segments which are flat transparent plastics segments with a radius corresponding to 80% of the van der Waals' radius of hydrogen [10]. These can be attached to a free rod or tube and so help to give an impression of space-filling characteristics. They are useful in determining the extent of interaction between nonbonded atoms.

Although only one scale is possible for most of the Dreiding models, hydrocarbon models of double the normal size can be made for demonstration purposes by means of the carbon and hydrogen extension units.

The problem encountered with the most precise space-filling models is even more apparent with the Dreiding type in that precision entails the provision of many units representing a range of chemical environments for a single element. The necessity for many units, especially if they are made in a durable material like stainless steel, involves a considerable outlay. While such an outlay can be justified for research tools, the Dreiding models are not likely to become extensively used as a teaching aid because of their high cost.

The relatively new 'Plastic Stereomodels' are similar in most respects to the Fieser models described below. The latter ceased to be commercially available some years ago but I had hoped that evidence of a demand for such models might have induced the manufacturers to start producing them again. Büchi now manufacture some models with plastic centres and rods but metal tubes (see Fieser models, 5.2.3 below) which can be cut, bent or shaped, the two last being achieved by heating the material. The linkage mechanism is the same as that for stainless steel Dreiding models (rod into tube) but the scale is 50 mm ≡ 0.1 nm (see Fieser). Double bonds are represented by a single unit with two nonrotatable links. There are three of these for carbon, nitrogen and oxygen double bonds. The other units are tetrahedral carbon and nitrogen and bivalent oxygen. Coloured spheres for attachment to the ends of tubes or rods are used to represent atoms or groups and are supplied in four colours. All the other units are appropriately colour-coded. Although there are only a few units, they are very readily adapted and can be used to represent a wide range of chemical species including rings containing four to seven atoms. Tetrahedra are supplied in two halves which must be glued together.

There is one basic set and five supplementary sets, all moderately priced.

A specific application of the Plastic Stereomodels is discussed in Chapter 8.

### 5.2.3 Fieser Stereomodels (Figure 5.4)

The Fieser models have been described as 'plastic Dreiding', and the principle of model construction is indeed similar. The scale is double that of the Dreiding models at 50 mm ≡ 0.1 nm. Each tetrahedral unit is formed from two rods and two tubes, the former of the plastics material ('Lexan', a polycarbonate) and the latter of aluminium. The extent of insertion of tube into rod is determined by indentations on the rods which fit into projections on the tubes' inner surfaces. Rotation about the linkage can occur but the friction is sufficient for a particular conformation to be retained. The carbon/carbon double bond is represented by a single unit with two rods joining two half tetrahedra indicating the $\pi$-contribution. This contrasts with the steel Dreiding models where, basically, the $\sigma$-bond framework is shown and $\pi$-character is only reflected in the shortening of the bond. To reduce cost, the carbon tetrahedra may be purchased in two halves to be cemented together. This is accomplished by painting each surface with methylene chloride before pressing together. Both trivalent (pyramidal) nitrogen and divalent oxygen units, with all angles tetrahedral and blue and red centres respectively, are available, together with polyethylene spheres in four sizes which may be used to represent substituents and terminal groups such as hydrogen if desired. These contain two holes, one of which can take a rod and the other a tube. Numbered pressure labels, an 'Angström rule' and Professor Fieser's book *Chemistry in Three Dimensions* [11] are accessories. There are three kits for student use, all inexpensive. The book is addressed to the student and if used in conjunction with the models would provide an excellent self-teaching method for introductory organic chemistry.

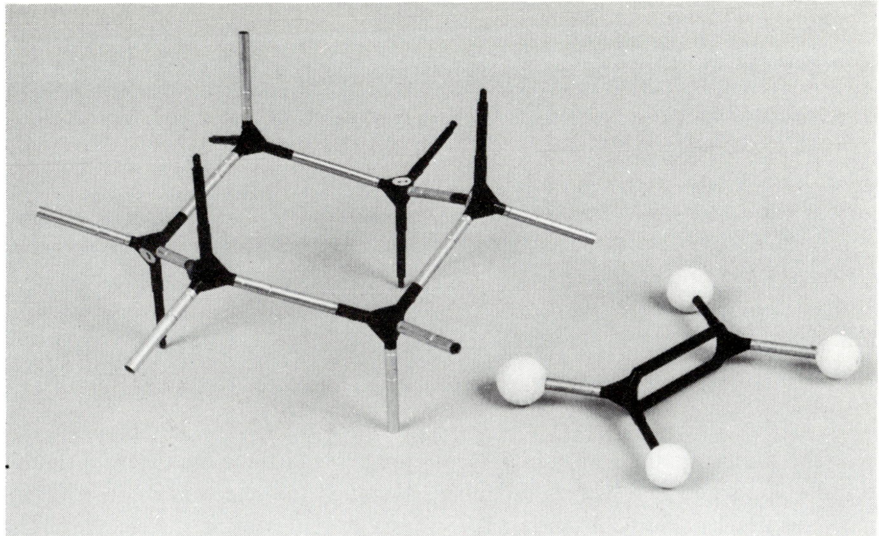

Fig. 5.4 – Fieser: (a) cyclohexane, (b) ethylene.

Although the number of units is small they can be readily adapted. Because the plastics material can be bent through 20-30° without permanent set, strained structures can be built. The rod can be permanently bent by softening with a hot-air blower before bending. Alternatively, rubber tubing may be used. Other atom units can be made from those in the set, as the plastic can be cut and painted with a quick drying enamel. These models are adaptable, geared to the teaching of fundamental organic chemistry and a great deal cheaper than the Dreiding ones. Applications and adaptations are discussed in two articles by Fieser, [12], [13] while Leska [14] describes a homemade construction aid for measuring dihedral angles on Fieser models.

### 5.2.4 Framework Molecular Models (FMM) (Figure 5.5)

Designed by Brumlik and developed by Brumlik, Barrett and Baumgarten [15] as a teaching aid for introductory organic chemistry courses, the FMM models have since found wider application (see below and Chapter 6). Plastic tubing which is similar to, but tougher, harder and less inclined to kink than plastic drinking 'straws', is joined by pronged metal units. Three of the pronged units, called 'valence clusters', have four, five and six prongs arranged pointing towards the corners of a tetrahedron, trigonal bipyramid and octahedron, and are silver-, brass- and copper-coloured respectively. These represent atom (or ion) centres. In addition there is an angle (90°) connector and a linear fastener tapered at either end.

Fig. 5.5 – FMM: (a) methane, (b) glycine, (c) ethylene.

The plastic tubing comes in 120 mm lengths and there are twelve different single colours and five two-colour tubes. Although the selection of colours permits the usual colour convention to be used in representing the elements, the manual and brochure propose a slightly different one. The bicoloured tubes are used to represent bonds between different atoms. For example, the C–H bond is black for carbon and white for hydrogen and consists of a black section corresponding to the covalent radius of carbon adjacent to a white one for the covalent radius of hydrogen. A black ring for the hydrogen nucleus follows and then there is a white section proportional to the van der Waals' radius of hydrogen. The O–H (red and white) and N–H (blue and white) tubes are arranged similarly. One 120 mm tube comprises two such series of 'bonds' and may be readily cut in two with a razor blade. Because the C–H, N–H and O–H 'bonds' are to scale (25 mm ≡ 0.1 nm), a small piece of each tube is left over. The scale is in fact the same as for the Dreiding models but a completed assembly looks somewhat larger because of the extra lengths of tube representing the van der Waals' radii. However, the two other bicoloured tubes have only two coloured sections per 'bond' (three for each 120 mm tube) representing in black/red the two covalent radii for the carbonyl bond and in black/blue those for multiple carbon/nitrogen bonds. To depict normal $\sigma$ bonds between different atoms, two valence clusters are linked by two tubes of different colour representing the covalent radii of each atom, which are themselves joined by a linear fastener. Although the tubes are flexible enough to build strained structures including small rings, they will kink irreversibly if bent through too great an angle. Over a period of years they become somewhat brittle and may break at the point reached by the tip of the metal prong inside the tube. On the whole, assemblies of these models are robust and can withstand repeated handling. They are easy to manipulate, no tools being required except that pliers may sometimes be necessary to remove linear fasteners because the metal units fit tightly into the tubes.

Apart from shortening internuclear distance, $\pi$-bonding can be indicated by forming extra links between the metal units. For example to construct a model of ethylene two trigonal bipyramidal units can be used, the equatorial prongs having tubing attached to represent the coplanar $\sigma$ bonds ($sp^2$ hybridisation) between carbon and hydrogen and between the two carbon atoms. Perpendicular to the plane of the C-H 'bonds' are constructed two rectangles of tubing representing the $\pi$-bond above and below the plane. These are formed by attaching vertical pieces to the axial prongs of the valence clusters and joining each vertical pair with a horizontal tube attached by angle fasteners (Fig. 5.5). Again, a model of acetylene is assembled by using octahedral metal units so that two pairs of rectangles can be constructed perpendicular to each other. Similar techniques can be used for all other multiple bonds. In every case restricted rotation about the bond results.

These cheap, versatile models are intended primarily for elementary teaching, but they have been successfully used in research where a less expensive substitute for Dreiding models was sought and a high level of precision was not essential. Apart from the student kit, all components may be purchased as packages containing one hundred similar units. Therefore models of macromolecules and extended arrays can be built with them. In teaching a course on inorganic crystal chemistry I have found these and similar models (see 5.2.10 below) invaluable for demonstrating the ways in which tetrahedra and octahedra may be linked to form chains, sheets and three-dimensional structures. Some examples are given in Chapter 6.

The FMM models can also be used to advantage in the study of optical isomerism in metal complexes. Chelate groups can be indicated by attaching an extra long tube, if necessary composed of several short ones joined by fasteners, to two adjacent positions on the valence cluster. This then simulates the usual graphical representation of a chelate group. Students much more readily accept the chirality of octahedral chelate complexes if they can handle models of the enantiomers and convince themselves that one is the mirror image of the other. This is even more true of those strange resolvable tetrahedral species with unsymmetrical chelate ligands which do not reveal their finer structural points when portrayed in two dimensions. Here a short and a long tube of different colours can form the chelate 'bow' and so convey the asymmetry. A simple, neat way of making FMM metallocene models is described by Sutton [16]. Ferrocene is a useful model to have when discussing symmetry properties because of its $S_{10}$ improper axis and dihedral angles, and this version does very well. A combination of Dreiding and FMM models is recommended by Chipman [8] for the construction of highly strained or 'twisted' species such as the triptycyl radical or trienes with *trans* double bonds. The more flexible FMM tubes are used for the most strained links. Although padding of the Dreiding rod with sticky tape where it is inserted into the FMM tube may be necessary, the Dreiding tube is large enough to fit tightly.

In order to extend the use of FMM models to molecules in which $\pi$-bonding involving d orbitals occurs, Barrett [17] designed a new valence cluster. Linear fasteners are inserted into a one inch diameter plastic sphere along the axes of the d orbital lobes. It is then possible to demonstrate $p_\pi$-$d_\pi$ bonding as in $Ni(CO)_4$, $d_\pi$-$d_\pi$ as in $Ni(PCl_3)_4$ and $\delta$ bonding, the example given being the hydrated copper (II) acetate $[Cu(OOCCH_3)_2]_2\ 2H_2O$, where the short Cu-Cu distance is attributed to lateral overlap of the copper $d_{x^2-y^2}$ orbitals. If many of these spherical valence clusters are used, the assembly begins to look like a ball and spoke model. In fact, for many purposes a Linnell plastic sphere (see Chapter 4) could be used, because the FMM tubes fit snugly into the holes in these. A similar principle has been employed in the design of the Unit models (Chapter 4.4) where pegs are inserted in the spheres in such a way that the common symmetries may be realised. The Linnett double quartet theory has been illustrated using FMM models [18], tetrahedral clusters and the tubing being used to show 'spin sets' of four, five, six and seven electrons. The completed octet and the electronic structures of diatomic molecules may also be demonstrated.

### 5.2.5 **Geodestix** (Figure 5.6)

As with the FMM models, only two types of component comprise the Geodestix system. These are coloured plastic rods of various lengths between one and twenty-four inches and soft, flexible plastic connectors. Six or nine inch aluminium rods are also available. However, in FMM models the tubes radiate from the valence clusters or connectors, which so impose a limitation. In Geodestix models the connectors form vertices of polyhedra and therefore there is no restriction in the type of polyhedron which can be constructed.

The connectors are flat, about one inch in diameter and with five, six or eight tubes (into which the rigid rods fit) radiating from the centre. The eight-tube connector can be obtained with either a central hole or a 'through-way' joining a pair of opposite tubes. Any scale may be used because the rods are easily cut with a knife or small hack-saw. In fact if the rod is nicked with a knife or ampoule file it may then be snapped between fingers and thumbs like a glass rod. A convenient scale for demonstration purposes is about 80 mm ≡ 0.1 nm. The connectors are so flexible that they will adopt virtually any orientation in an assembly, usually without permanent deformation. However, they should be carefully eased off the rods when dismantling to avoid tearing. Because they are cheap and easily replaced they are the sort of model which can be lent to students for private study. I have frequently handed out polyhedra to members of a group of students and asked them to work out their symmetry properties. These are later discussed in a tutorial. It is not unusual for the models to undergo metamorphoses in the hands of the students, but little is lost and much gained!

In common with the FMM models, Geodestix can be used to demonstrate isomerism in inorganic chemistry, chelate groups being represented by longer rods bowing outwards from an edge of a polyhedron and attached to the same connectors as the rod forming the edge. Very short coloured rods can be used as markers to identify particular vertices by inserting them through the central holes of the connectors. This provides a simple demonstration of how there can be no *cis-trans* isomerism in unidentate tetrahedral complexes. Take a model of a tetrahedron, insert markers in two corners and ask the students whether any other arrangements are possible. Similarly, it is possible to draw on the board a number of apparent isomers of a chelate octahedral complex which *look* different. The students are then left to find out from the models that some are actually the same structure.

At a more advanced level, Geodestix models have been used [19] to demonstrate borane structures [20]. Starting with a tetrahedron, connectors representing boron atoms are added one at a time through all co-ordination numbers up to twelve (the icosahedron). It is shown that for some, more than one symmetry may be possible. The models can be used to illustrate similar structural sequences in the carboranes. Geodestix are particularly suitable for the portrayal of unusual co-ordination numbers and geometries. They are also suitable for showing the relationship between pairs of polyhedra such as the pentagonal bipyramid and the monocapped octahedron, as Muetterties and Wright [21] demonstrated with ball-and-spoke models (Chapter 4.1). It has been pointed out [22] that some apparently rather unsymmetrical co-ordination numbers are related to simpler symmetrical ones. For example, joining five edge centres in a ten co-ordinate structure could form a trigonal bipyramid. I have been able to show that this is true for the eight-co-ordinate trigonal dodecahedron by making a model of this structure with an included tetrahedron. Spokes from the central sphere are tetrahedrally arranged and point to four edge centres (Figure 5.6).

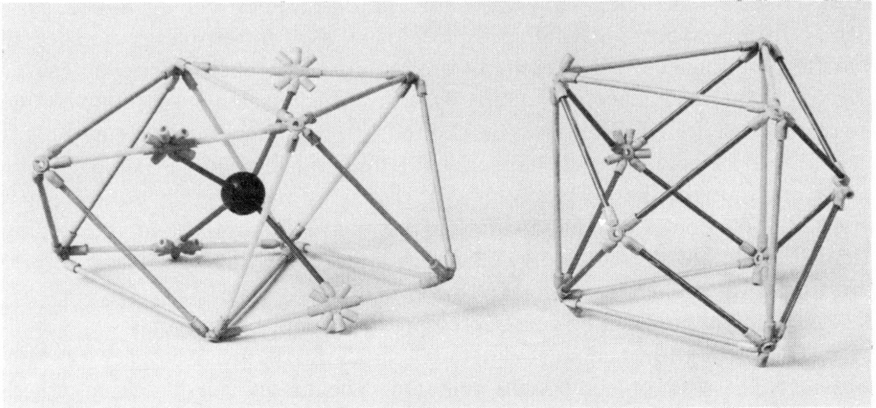

Fig. 5.6 – Geodestix: (a) trigonal dodecahedron (bisdisphenoid) with included tetrahedron, (b) tricapped trigonal prism.

Very similar models are marketed in the USA under the name 'Think Sticks'. The connectors are of the same type but the rods may be of plastic or wood with lengths between two and twelve inches. Long lengths of stick may also be purchased and cut to size. Until recently the same company, Edmund Scientific, also produced a kit called 'Super Structures' in which the star-shaped connector unit had twenty-six spokes over which plastic tubes fitted. Variously shaped panels for covering sections of completed polyhedra were included.

The Wells kit from Oxford University Press contains a variety of models, some of them of the Geodestix type. There are twelve-, ten- and eight-inch rods and five- and six-way connectors.

### 5.2.6 Geoframe (Figure 5.7)

This skeletal model kit from Griffin and George is similar in principle to Geodestix in that it is intended for the construction of any polyhedron. Square section tubes with an outside edge about 10 mm across are joined by flat 'vertex pieces' having three, four, five or six prongs. Near each end are two holes diagonally opposite each other and when a prong of a 'vertex piece' is inserted into the tube a projection on the prong fits into one of the holes. Insertion or removal is facilitated by squeezing the tube so as to widen temporarily the gap between the holes. For construction of a polyhedron where more than six edges meet at one corner, two vertex pieces may be joined together by means of a 2BA nut and bolt included in the kit. This provides eight-, ten- or twelve-pronged units as required. In common with the Geodestix connectors, the 'vertex pieces' are flexible.

The tubes are available in two lengths so that when the model is assembled the distance between the 'vertex pieces' is either 200 mm or 400 mm. The latter tube length produces very large models and the shorter length is adequate for most purposes including lecture demonstrations. Tubes can be joined by a tightly fitting, appropriately shaped piece of wood if longer pieces are needed. On the other hand, if shorter lengths are required the tubes as supplied can be cut and extra holes made near the ends with a small hacksaw. Alternatively the holes may be drilled. The hacksaw method is quick, and if only a tiny hole is made it can be enlarged and shaped by the tip of a ball-point pen. Care must be taken not to make the holes too large or the tubes may split when the prong is inserted. The completed polyhedra can be converted into 'solid' models by attaching suitably shaped card to each face either with sticky tape or with wire threaded through the central holes in the connectors.

Any polyhedron can be constructed providing enough components are available. The instruction pamphlet illustrates about forty, giving in each case the number and type of component required. The assembled models are rigid, considerably more so than the Geodestix ones, and robust. The tubes are white and thick and therefore show up well in a large room. These are cheap, large-scale models. Assembly is a little hard on the fingers but no tools are required and the resulting polyhedra are pleasing.

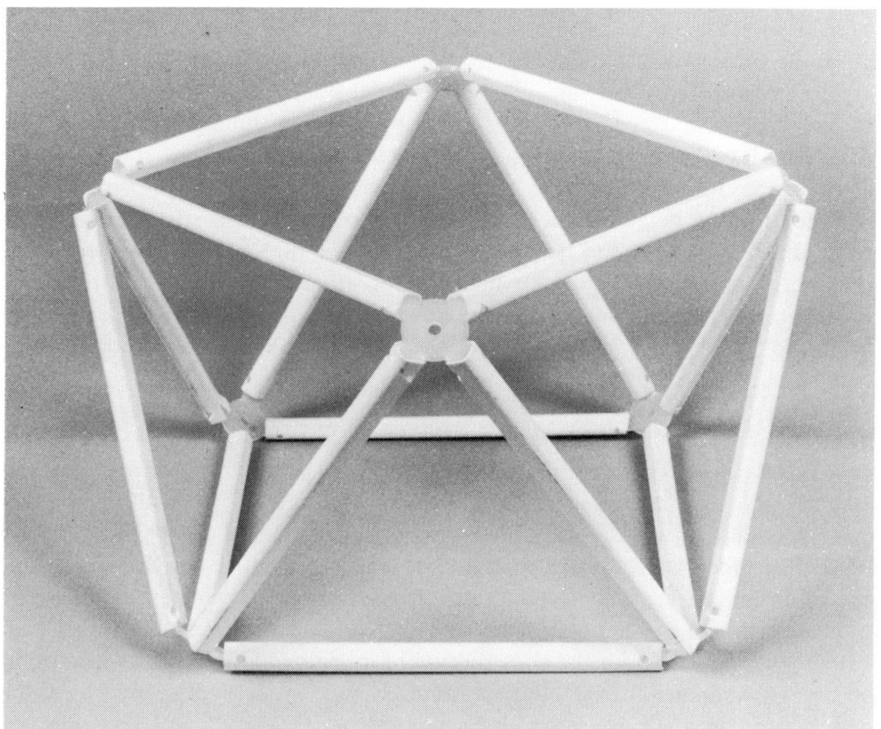

Fig. 5.7 – Geoframe: square antiprism.

### 5.2.7 Godfrey Stereomodels

About two hundred components with twenty-one different species representing nine elements comprise this fairly simple kit designed by Godfrey [23] as an aid to teaching organic chemistry. Flat pieces (2 mm thick) of low-density polyethylene form atom species, groups or bonds and these are connected by pieces of transparent tubing which 'snap-lock' into position holding the joined pieces an appropriate distance apart. The scale is 25 mm ≡ 0.1 nm. Tetrahedral carbon and nitrogen models are each formed from two flat pieces of polyethylene both of which have two prongs and a slot. The two parts are slotted together at right angles to each other so that the prongs are tetrahedrally disposed (see Disc models, 4.4.6). Divalent oxygen and sulphur consist of one such piece. Rectangular or capsule shaped pieces are used for the halogens and multiple bonds, olefinic, aromatic and acetylenic, all with the necessary number of prongs for joining. Models of benzene and cyclopropane can be supplied because assembled units and amide and nitrile groups are each represented by one unit.

It might appear from the above description that the models are ball-and-spoke/skeletal hybrids but in fact the emphasis is very much on the bonds. In a completed molecular model, the six nontubular atom models, with the exception of sulphur, are small and not very obvious. The types of multiple bond already mentioned, the halogens, hydrogen, carbonyl oxygen, thionyl sulphur, amide and nitrile groups are all represented by lengths of tube proportional to the appropriate covalent radii. The tips of these tubes are colour-coded. The van der Waals' envelope is not indicated in these models. Free rotation about single bonds is possible but the friction is sufficient for any desired configuration to be retained. The 'snap-lock' connection is sufficiently strong for strained structures and large assemblies to be built. No tools are needed and the units are easily adapted to make other atoms or groups. Godfrey [23] describes the preparation of an isocyanate unit and also the use of additional tubing for hydrogen bonds. The models are sold as a kit but extra components may be purchased separately.

### 5.2.8 Kendrew Skeletal Models

These consist of units of rigid brass rod which are joined by insertion of a rod on each unit into a small barrel bearing two screws. When the screws are tightened they fit into grooves on the rods and the whole assembly is securely held. The models are precise with a scale of 20 mm ≡ 0.1 nm. Orientation of the units is accurately achieved by use of angle gauges, seventeen of which are available with angles between 100° and 130°. The units are highly specialised, being designed solely for the construction of models of biochemically important species. A well-known example of a Kendrew-type model is that used by Crick and Watson in their work on the structure of DNA [24]. All the units are designed for the construction of models of complex proteins and similar molecules and could not readily be used for other purposes. Amino-acids and base pairs are supplied as single units as are specially designed oxygen, sulphur and phosphorus species. Wire is used for hydrogen bonds. The components are all sold separately and also as kits. Also now available are eight kits of components for the construction of models of a bacterial cell wall (in two versions, one with 'peptide tails'), of DNA, RNA, chymotrypsin, lysosyme, myoglobin and ribonuclease. Two metal frames (large and small) can be supplied to support the models and the grids forming the top and bottom of these frames help with the orientation of the model during assembly. Calipers calibrated so that angström distances can be measured on the model may also be obtained. The two pairs of cursors permit both external and internal measurements to be made.

### 5.2.9 The Kennard-Doré Stereo-Model Construction Set

This sophisticated kit is designed for the preparation of precise permanent models by soldering together pieces of metal rod. As well as a soldering iron, flux, brazing alloy, guillotine and rod, the components provide a means of holding the rods which are to be soldered together in the correct orientation. The rods are cut to size to within 0.5 mm with the adjustable guillotine. Two systems are available for supporting the rods which are to be joined. The first employs magnetic chucks to which are attached vertical support rods bearing clamps. These can be moved to the required position on a metal base plate and then locked in place. Fine vertical adjustment is achieved by a milled head screw with 10 mm travel. Thirty-six plastic angle templates with angles between 55° and 125.5° are included in the kit. Having selected one of these, the model-builder moves the vertical holders on their magnetic chucks until the rods to be joined touch each side of the angle piece. The holders can then be locked in position while the soldering is done. A small pendant clinometer is provided to adjust angles where the junction between rods is nonplanar.

An alternative means of assembly makes use of ring magnets to which adjustable sleeve vertical supports are attached. From a knowledge of atomic co-ordinates these can be used to position the atomic centres precisely in space. Using magnetic chuck supports, metal rods are then held between the atomic centres and soldered together. Special metal sleeves are used where rotation about a bond is possible and plastic sleeves supplied for helical structures where greater flexibilty is desirable.

The Kennard-Doré kit is certainly expensive. However, it provides a means of making precise, permanent models to any scale and according to the user's own specifications where suitable commercial models are not available or may be even more expensive.

### 5.2.10 Minit and Orbit Models (Figures 5.8 and 5.1)

Though based on the same principles of construction as the FMM models (see 5.2.4 above) the Minit and Orbit components are all of plastic and the 'atom centres' are very much more varied. There are fifteen different 'atom centre' shapes in the Orbit system and nineteen in the Minit system, the extra four being two-co-ordinate angular oxygen species. The number of prongs on each 'atom centre' ranges from one to six and there are also eight- and twelve-co-ordinate arrangements. Symmetries are linear, angular (various angles), trigonal planar, square planar, tetrahedral, trigonal bipyramidal, octahedral, cubic and twelve-co-ordinate (for cubic or hexagonal close-packed structures). The twelve-co-ordinate 'atom centre' is in three parts. Two sets of three groups can be connected either side of a planar six-pronged piece and whether the symmetry is that of cubic or hexagonal close-packing then depends on the orientation of the two three-pronged pieces with respect to each other. When coincident, a hexagonal, close-packed unit results, when 'staggered', a cubic close-packed arrangement. Fourteen different colours are used and a total of fifty-nine different Minit 'atom centres' (Orbit fifty-five) are manufactured which differ in either colour or shape. The usual colour convention is used. Bond angles are identified on 'atom centres' by code marks where there may be ambiguity. There are five planar three-pronged types, all of which are available in black (for carbon) while three are available in blue (for nitrogen). The angles between the three prongs in these species are such that it is possible to construct a variety of cyclic systems including strained, fused and heterocyclic rings. All the components necessary to build models of macromolecules such as proteins, polysaccharides, DNA and RNA are produced, among them a hydrogen bond model and a peptide link. All the common symmetries occurring in inorganic crystal structures are represented so that three-dimensional arrays of MX, $MX_2$ and $MX_3$ structures are easily assembled (Figure 5.8).

The atom centres are joined by plastic 'straws' which are either precut or cut to size by the user. The chief difference between Minit and Orbit models is that the scale of the former is smaller, 20 mm ≡ 0.1 nm being the recommended one, although 12.5 mm ≡ 0.1 nm is possible because the atom centres themselves are smaller than those of the Orbit system. The recommended scale for the Orbit models is 30 mm ≡ 0.1 nm. Of course, the user can vary the scale, but where the

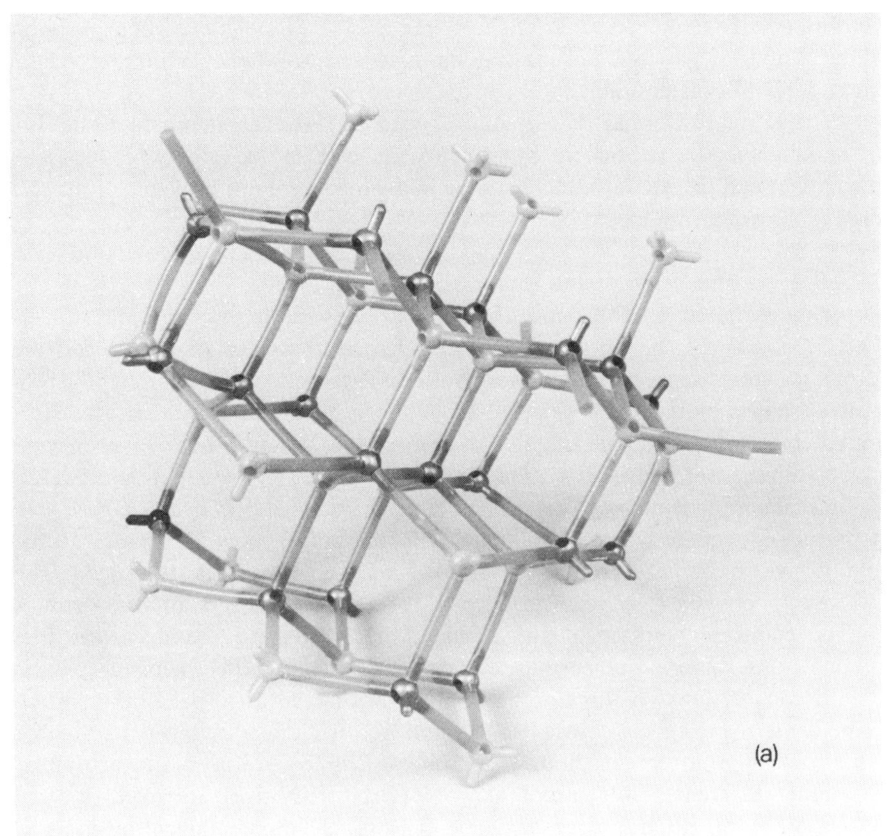

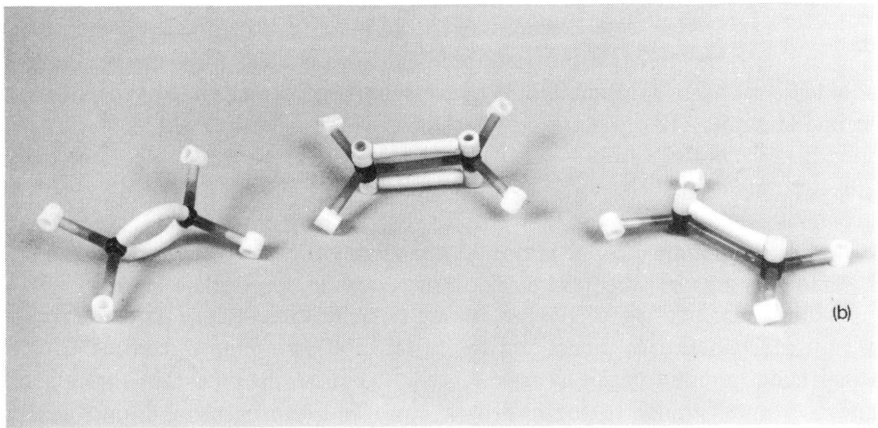

Fig. 5.8 – Minit and Orbit: (a) Minit wurtzite (ZnS), (b) (i), (ii) and (iii) Minit versions of ethylene, (c) Orbit fluorite ($CaF_2$).

rods are precut the scale is according to the manufacturer's recommendations. The tubing is mostly rigid but some flexible is supplied and this can be used for strained structures, multiple bonds or for simulating orbital shapes, chelates and so on. The rigid tubing (mostly green, but some red or white) will kink and even split if not handled carefully or if the fit of the tube on the prong is too tight and too much force has to be used. However, the fit is good generally and few problems are encountered.

Multiple bonding can be indicated in a variety of ways, the simplest being to shorten the tube between the atomic centres by an appropriate amount. Other possibilities include use of flexible tubing (i) as in the ethylene model shown in Figure 5.8, where two pieces join the tetrahedral carbon atom centres. On the other hand 'pi-bond pegs' (ii) can be used. These consist of a right-angled piece and two units similar to the hydrogen atom and hydrogen bond units except that both the single-pronged and two-pronged (linear) unit have a peg at right-angles to the prong or prongs instead of a hole. These are able to fit into the hole in another atom centre. In the ethylene model (Figure 5.8) the single-pronged species support a piece of flexible tubing parallel to the rigid tubing representing the sigma bond. This is unsatisfactory because it suggests that the pi-electron density is all on one side of the plane of the C–H bonds. A better system (iii) is to use hydrogen atom centres on either side of the rigid tube to support two pieces of flexible tube parallel to each other and the central tube.

Yet another possibility is to use the right-angled pieces to form rectangles above and below the plane of the C–H bonds as with FMM models (Figure 5.5). For aromatic systems either single and double bonds may be alternated or trigonal carbon atom centres used and the tube shortened. Through the holes in the atom centres extra tubes may be inserted perpendicular to the plane of the ring and then joined using the angle pieces, so simulating the $\pi$-clouds above and below the ring. Rotation about single bonds is free but there is sufficient friction for configurations to be retained. Rotation is restricted in all cases where more than one tube is used to join atom centres. As the manufacturers point out, the atom centres and tubes can also be used to demonstrate both how the lobes are distributed in atomic orbitals as well as hybridisation in molecular orbitals. Of course, the tubes can only indicate the direction of the lobes, not their contours.

Space-filling shells which are not available for the older Orbit type have been designed for use with the relatively new Minit models. The skeletal model is first made up then cladding is accomplished by fitting the shells over the atom centres. Each shell is made in two halves, one containing a peg and the other a hole into which the peg can fit. The exception is the tetrahedral carbon shell, which has a peg and a hole in each half. The uni-, bi- and tri-valent atom centre types all have their prong or prongs attached to a ring which has a hole right through the centre. All but one of the shells are attached to the skeletal model by passing the shell peg through this hole in the atom centre. However, the tetrahedral carbon atom centre does not have a central hole and so two pegs are necessary to hold the shell securely in position.

There are seven different shells. They represent hydrogen, hydrogen-bonding hydrogen (flatter), tetrahedral and trigonal carbon, bivalent and carbonyl oxygen and bromine. The scale is 20 mm ≡ 0.1 nm as recommended for Minit skeletal models. It is possible to construct simple space-filling models using these shells without any internal framework (Figure 5.9) by means of shell links. These are plastic connectors each with two constrictions around which adjacent shell links can be fitted. They are quite satisfactory but limited by the number of shell types available. Perhaps the most helpful application of the shells is to clad only those parts of a skeletal model where steric hindrance might be significant, as in *ortho*-substituted ring systems. A simple example is shown in Figure 5.9. Unfortunately, although light the shells are substantially heavier than the skeletal part of the model and tend to pull it apart unless it is carefully supported.

Fig. 5.9 – Minit with shells: (a) acetic acid, (b) *o*-chlorotoluene.

Sixteen fully assembled models of metal crystals and important inorganic compounds can be purchased. Some of these are in the Orbit and some in the Minit system. An assembled Minit model of DNA is also listed. Specific kits of components for each of these can be obtained at a price which is, on average, half that of the assembled model.

Various kits are produced. Organic/inorganic chemistry, lattices and biochemistry kits are supplied in both Orbit and Minit. In addition there is an Orbit demonstration set and one on basic structures. Each of the student kits includes a booklet which can be used for self-instruction. The Minit booklet for the organic/inorganic chemistry kit is generally good though inevitably containing some errors, mostly printing or minor. However, contrary to the statement in the text, it *is* possible to construct isomers of the complex ion $[Co(en)_3]^{3+}$, as Figure 5.10 shows! In addition to a booklet, work cards are issued with the Orbit kits and with assembled models giving essential structural data.

Fig. 5.10 – Minit: optical isomers of a tris(ethylenediamine) octahedral metal complex.

The Orbit or Minit systems together with their instructions provide a satisfactory and enjoyable way of learning basic structural chemistry. Moreover the systems can be extended to any desired degree of complexity and used effectively at advanced levels of chemistry teaching.

A kit of Orbit models and some cutout molecular shapes are included in Pearce and Glynn's programmed learning package (text plus models) *Stereochemistry: An Introductory Programme with Models* [25]. The book begins with a brief discussion of the various types of molecular model and the programme which follows illustrates their utility in the teaching of organic stereochemistry.

Colour-Coded Skeletal (CCS) Models initially proposed by Stong [26] and developed by Clarke [27] comprise Minit atom centres with colour-coded plastic tube connectors plus colour-coded plastic sleeves which are used to identify particular atomic species. Crimped metal sleeves (also colour-coded) may also be used to prevent rotation or to prevent tubes from slipping off prongs in strained systems. If the recommended scale of 12.5 $\equiv$ 0.1 nm is used, the models can be used in conjunction with CPK models (see Chapter 2.3.2). A partial space-filling effect may be brought about by attaching polystyrene spheres to terminal tubes. These spheres, most frequently representing hydrogen or carbonyl oxygen, give an overall impression of the van der Waals' envelope but the internal bond geometry of the molecule is still visible (see Atomunits, 5.2.1 above, where segments rather than whole spheres are used for partial cladding). Now commercially available, these models have been used to study conformations of small molecules and enzyme-substrate interactions [28]. They are currently marketed as an introductory set or as components.

Lab-Aids have recently produced four simple kits similar in design to the Minit/Orbit system. These are the 'Molecule Model', 'Organic Chemistry (Hydrocarbons)', 'Organic Chemistry (Functional Groups)' and the 'Isomers' kits. Each contains twelve small sets for individual student use, worksheets, instructions and background information for teachers. They are suitable for chemistry teaching at an elementary level.

Various constructional toys for children such as 'Construct-O-Straws' and 'Flexostraws' are based on similar principles, the former actually being manufactured by the same Company as Minit and Orbit (RJM Exports Ltd.). These may sometimes be used as substitutes for model building kits but they usually have severe limitations. All the 'Construct-O-Straws' connecting pieces (joined by plastic 'straws') are rigid and planar while the 'Flexostraws' kit has an octahedral piece but no tetrahedral unit. With the advent of simple, inexpensive skeletal models of the Minit and Orbit type, there seems to be little point in spending time adapting rather unsuitable models unless they have been discarded by their original owners.

## 5.3 MISCELLANEOUS AND DO-IT-YOURSELF (DIY) SKELETAL MODELS

Simple framework models can be made by bending wire into suitable shapes and there are many examples of this type of model in the chemical education literature. It is possible to buy wire, such as the 'Fieldhouse modelling wire', designed specifically for this purpose, but some of suitable gauge and flexibility can usually be found in most workshops. Variants on the bent wire theme are numerous. Doré [1] clad his bent wire models with segments of soft polyurethane foam glued on with 'Evostick' so as to indicate the extent of the van der Waals' radii of the constituent atoms. This principle was subsequently developed into 'Atomunits' and 'mushrooms', see 5.2.1 above. Though brief mention of some skeletal models of macromolecules follows in order to indicate the types of materials used, these are discussed in detail as assemblies in Chapter 6.

Vedvick and Coates [29] made a bent wire model of haemoglobin in which helical segments were attached to a larger gauge backbone. The haem units were of brass sheet and soldered in position. Kaye [30] made polymer models from plastic-coated clothes lines with pipe-cleaner side-chains, while Nicholson [31] used coloured pipe-cleaners exclusively. Each pipe-cleaner colour represented a particular chain segment in a macromolecule such as an enzyme or a protein. The pipe-cleaners were joined by twisting the ends together. Dreiding-type units of vinyl covered wire were devised by Larson [32]. For example, a tetrahedral unit could be made by twisting two pieces of vinyl covered wire together and then fusing the vinyl sleeves at the junction by application of the hot tip of a soldering iron. Units can be joined either by bending the ends and hooking them together having previously retracted the plastic sleeving, or by inserting the wire of one unit into the vinyl sleeve of the other. Stability of the assembly is improved if vinyl cement is applied. Joining may also be accomplished by melting the plastic just inside each piece of sleeve with a soldering iron immediately before the wires are inserted. Coloured sleeves may be used to identify particular radii and plastic foam attached to indicate van der Waals' radii and electron density due to $\pi$-bonding.

Molecular models built from aluminium tubing were used by Harris and Cheung [33] in their study of transition states in some configurational inversions. The models indicated where overcrowding resulted in molecular deformation and it was possible then to infer which reaction paths were most probable. The tubing is sawn to appropriate lengths for a scale of 38 mm $\equiv$ 0.1 nm allowing for connector lengths. Connection is made by lining the ends of each tube with nylon sleeving. The narrow end of a brass soldering tag is then forced into the end of each tube where it is securely held by the nylon sleeve. Two units are joined by a nut and bolt using the holes in the soldering tags. In-plane and out-of-plane angular deformation can be indicated by, in the former case, adjusting the arrangement of the bonds before final tightening of the nuts and bolts and, in the latter case, by bending the soldering tags. Where it is necessary to demonstrate free rotation, the bond unit is sawn in half and a brass rod covered with a nylon sleeve inserted which then acts as a pivot.

Several homemade models of the Kendrew-type employing brass rod have been reported, but as these are models of macromolecules they are discussed in the next Chapter.

In the teaching of structural inorganic chemistry it is often helpful to compare framework with other models or to use 'mixed' models with some space-filling characteristics. Dunstone [34] devised a project in which model-building was used as an instruction medium in the study of silicate structures. Framework models,which he believed to be the most satisfactory for this purpose, were compared with space-filling and ball-and-spoke models of the same structures. Hicks [35] discusses both 'tangential-sphere' models of inorganic structures and clad framework models, while a 'mixed' model of a zeolite framework, partially space-filled, has been used in conjunction with space-filling models of sorbate molecules to the same scale (see Chapter 6.3.1) [36].

It will be apparent that skeletal models are used very extensively and have been applied to many aspects of chemistry. As teaching instruments they have enormous versatility and have the great advantage of being available at low cost. They are also easy to manipulate. On the research side, the precise framework models are indispensable in some structural studies [37] and can be used quantitatively.

**Table 5.1**
Skeletal models

| Model and/or supplier | Type available | Price code (Table 1.4) |
|---|---|---|
| Atomunits | k, s/f, c | a, b |
| Dreiding | k, s/f, c | b, c, d |
| Fieser | k, c | a |
| Framework Molecular Models (FMM) | k, c | a |
| Geodestix | c | a |
| Geoframe | k | a |
| Godfrey Stereomodels | k, c | b |
| Kendrew Skeletal | k, c | c, d |
| Kennard-Doré Stereo-Model Construction Set | k | d |
| Minit and Orbit | p, k, s/f, c | a, b |
| Colour Coded Skeletal (CCS) | k, s/f, c | a, b |
| Lab Aids | k | a, b |

**Key** p–preassembled
k–kit
s/f–space-filling facility
c–components

*Note:* For Nicholson models see Chapter 6.2.1.

## REFERENCES


[1] Doré, C. F., *Educ. in Chem.*, **2**, 192 (1965).
[2] Doré, C. F., *Educ. in Chem.*, **7**, 194 (1970).
[3] Smith, I., Smith, M. J. and Doré, C. F., Chapter 40 in *Chromatographic and Electrophoretic Techniques,* **1**, edited by I. Smith, Heinemann (1969).
[4] Dreiding, A. S., *Helv. Chim. Acta,* **42**, 1339 (1959).
[5] Robinson, D. L. and Theobald, D. W., *Q. Rev. Chem. Soc.*, **21**, 316 (1967).
[6] Collins, L. J. and Kirk, D. N., *Tetrahedron Lett.*, **18**, 1547 (1970).
[7] McEachern, D. M. and Lehmann, P. A., *J. Chem. Educ.*, **47**, 389 (1970).
[8] Chipman, W. B., *J. Chem. Educ.*, **46**, 118 (1969).

[9] (a) ApSimon, J. W., Demarco, P. V. and Raffler, A., *Chemy. Ind.*, 1792, (1966).
(b) ApSimon, J. W., *Can. J. Chem.*, **46**, 808 (1968).
(c) ApSimon, J. W., Craig, W. G., Demayo, A. and Raffler, A. A., *Can. J. Chem.*, **46**, 809 (1968).
[10] Kooyman, E. C., *J. Chem. Educ.*, **40**, 204, (1963).
[11] Fieser, L. F., *Chemistry in Three Dimensions*, Rinco Instrument Co. (1963).
[12] Fieser, L. F., *J. Chem. Educ.*, **40**, 457 (1963).
[13] Fieser, L. F., *J. Chem. Educ.*, **42**, 408 (1965).
[14] Leska, J., *J. Chem. Educ.*, **50**, 516 (1973).
[15] Brumlik, G. C., Barrett, E. J. and Baumgarten, R. L., *J. Chem. Educ.*, **41**, 221 (1964).
[16] Sutton, J. R., *J. Chem. Educ.*, **47**, 305 (1970).
[17] Barrett, E. J., *J. Chem. Educ.*, **44**, 146 (1967).
[18] Baumgardner, C. L. and Wahl, G. H., *J. Chem. Educ.*, **45**, 347 (1968).
[19] (a) Greenwood, N. N., lecture-demonstrations on boranes and carboranes;
(b) Ellis, I. A., personal communication.
[20] For examples of these structures see Wade, K., *Chem. in Br.*, **11**, 177 (1975).
[21] Muetterties, E. L. and Wright, C. M., *Q. Rev. Chem. Soc.*, **21**, 109 (1967).
[22] See, for example, (a) Cotton, F. A. and Bergman, J. G., *J. Am. Chem. Soc.*, **86**, 2941 (1964);
(b) Al-Karaghouli, A. R. and Wood, J. S., *Chem. Commun.*, 135 (1970).
(c) Toogood, G. E. and Chieh, Chung, *Can. J. Chem.*, **53**, 831 (1975).
[23] Godfrey, J. C., *J. Chem. Educ.*, **42**, 404 (1965).
[24] Watson, J. D., *The Double Helix*, Weidenfeld and Nicolson (1968).
[25] Pearce, J. and Glynn, E., *Stereochemistry: An Introductory Programme with Models*, Wiley, (1976).
[26] Stong, C. L., *Sci. Amer.*, **234**, 124 (1976).
[27] Clarke, F. H., Jr., U.S. Pat. 3,939,581, Feb. 24, 1976.
[28] Clarke, F. H., *J. Chem. Educ.*, **54**, 230 (1977).
[29] Vedvick, T. and Coates, M. *J. Chem. Educ.*, **48**, 537 (1971).
[30] Kaye, H., *J. Chem. Educ.*, **48**, 201 (1971).
[31] Nicholson, I., *J. Chem. Educ.*, **46**, 671 (1969).
[32] Larson, G. O., *J. Chem. Educ.*, **41**, 219 (1964).
[33] Harris, M. M. and Cheung, K. L., *Chemy. Ind.*, 1378 (1962).
[34] Dunstone, J. R., *J. Geol. Educ.*, **20**, 88 (1972).
[35] Hicks, F. G., *J. Chem. Educ.*, **51**, 29 (1974).
[36] Walton, A., *Educ. in Chem.*, **9**, 146 (1972).
[37] Kirk, D. N., personal communication.

# 6

# Macromolecules, extended arrays and polyhedra

## 6.1 INTRODUCTION

The possibility of building models of macromolecules and extended arrays has been deliberately under-emphasised in Chapters 2, 4 and 5 because the purpose of these is generally to describe models of fairly wide application. However, many systems of the space-filling molecular, ball-and-spoke and skeletal types can be applied to the construction of macromolecules and some of these are listed in Table 6.1, with appropriate cross-references. Discussion in this Chapter is consequently limited to a consideration of their application in the macromolecular context. Similarly, some of the models described earlier can be used to build extended arrays and so only this aspect is emphasised in the ensuing discussion.

Models of a different kind from those previously mentioned, both those commercially available and homemade varieties, are also considered. The most interesting feature of some is a departure from the usual practice of representing every atom or bond by a single model unit. Instead a model unit may represent a large group of atoms. In this way it is possible to reduce size, cost and complexity.

The huge burgeoning of research into molecular biology in the last few decades has led to a more general interest in the structure of naturally-occurring macromolecules. This has spread through tertiary into secondary education, where study of these macromolecules is now a feature of many biology and chemistry courses in the VIth form. As a result, a number of students of biology, chemistry and environmental science already have some knowledge of these structures (DNA, inevitably, being 'top-of-the-pops') on entry to University. Follow-up courses are essentially of macromolecular chemistry and deal with the nucleic acids, proteins, polysaccharides and similar topics, tending to cut across conventional departmental boundaries. Such faculty-based courses are now possible owing to the existence of 'course unit' systems.

In addition to biology and environmental science, chemistry impinges on many other sciences. Among these, materials science is now accepted as an appropriate subject for study in schools. From this new development has grown a widened interest in natural and synthetic polymers and so a need for models of these structures.

Further 'blurring at the edges' has occurred between inorganic chemistry and crystallography, mineralogy, geochemistry and related subjects. This has resulted in an increased need for models depicting natural minerals, particularly the silicates, and new synthetic materials such as the 'molecular sieves'. The kind of model which is frequently helpful in these studies is one where it is possible to join single units so that they extend indefinitely in one-, two- or three-dimensions giving rise to chains, sheets or three-dimensional arrays. Polyhedra and the ways in which they may be linked to produce such structures, together with some specific applications in the teaching of structural inorganic chemistry, are discussed in this Chapter.

**Table 6.1**

Models discussed in Chapters 2, 4 and 5 which are suitable for macromolecular structures.

| Category | Type | Cross-reference |
|---|---|---|
| Space-filling | Courtauld† | Chapter 2.3.1 |
| | CPK + Boyd‡ | 2.3.2 |
| | Leybold† | 2.2.5 |
| | Molecular Fragments | 2.2.6 |
| | Scale Atoms | 2.3.3 |
| Ball-and-Spoke | Beevers‡ | Chapter 4.2.1 and 4.4.3 |
| | CSL‡ | 4.2.2 |
| | Molymod‡ | 4.2.6 and 4.4.3 |
| | SASM‡ | 4.2.4 and 4.4.3 |
| | SRM‡ | 4.2.5 |
| Skeletal | Atomunits† | Chapter 5.2.1 |
| | Dreiding† | 5.2.2 |
| | Kendrew† | 5.2.8 |
| | Kennard-Doré Construction Kit† | 5.2.9 |
| | Minit/Orbit‡ | 5.2.10 |
| | CCS | 5.2.10 |

†Designed primarily for building models of biochemical structures. The others have the necessary components but are also of wider application.
‡Including assembled models.

There is great diversity in the devices used to model macromolecules and teachers and research workers have shown considerable ingenuity in the use of a variety of materials and techniques. Abandonment of the 'atom-by-atom' approach to model building opens up all sorts of other possibilities for highlighting particular structural features. On the inorganic side a technique which does not seem to have been applied extensively as yet is the use of mixed models combining space-filling and skeletal characteristics as in molecular/ionic species in surface studies.

## 6.2 METHODS FOR MODELLING MACROMOLECULES

### 6.2.1 Application of Conventional Models

Table 6.1 gives cross-references to previous Chapters for general descriptions of types which have been or can be used for the construction of models of macromolecules. The criterion in most cases is that the kinds of component necessary to make a DNA model are obtainable without requiring major modification.

Of the space-filling variety, Ealing now market an assembled DNA from the excellent, but expensive, CPK models. A CPK model of elastin was used by Gray, Sandberg and Foster [1] to help elucidate the structure and function of this material. Assembly time can be saved with Edmund's 'pregrouped' models of atom clusters† which can be joined to form macromolecular species such as polypeptide models. Models of this type were designed and used by Landé [2] in his studies of peptide conformations and the effect of non-bonded interactions. It is claimed that these models effectively indicate steric hindrance. Moulded clusters of white polystyrene are joined by stainless steel pins or sockets and the faces of bonded atom models touch, making these Stuart-type models. The scale is small, being 5 mm ≡ 0.1 nm. Models of several macromolecules have been made, including insulin. Various macromolecules can be modelled from 'Molecular Fragments' and purchased (from SRM) already assembled. These include DNA, the silk molecule, polypeptides and amino acids.

Among manufactured ball-and-spoke kits, the Beevers models lend themselves particularly well to the construction of large assemblies because the scale is small. A photograph of a myoglobin model appears in an article by Beevers [3] and an alpha-helix model may be purchased already assembled. The CSL range includes a considerable number of macromolecules and related species, among them being assembled models of DNA, collagen, insulin, rubber, β-gutta percha, an alpha-helix and a set of twenty amino acids. Assembled 'Molymod' models of several of the commercially important synthetic polymers are available and also of DNA, an alpha-helix, a polypeptide, amino acids and cellulose. All these models can also be obtained as specific kits. A ball-and-spoke DNA model marketed by SRM has all the base pairs and sugar phosphate units preassembled but the purchaser has to mount them on a stand. An almost completely assembled alpha-helix model consists of twenty peptide units.

The DNA components can be obtained as a kit in which the appropriately

†Not in the latest Edmund catalogue, but may still be obtainable.

coloured polystyrene spheres have bonding positions marked for punching with the two punches provided. The spheres can then be joined either by glueing the wooden spokes into the smaller punched holes or by insertion of a 'vinyl receptor' (see Chapter 4.2.5) into a larger hole made by the larger punch. This can then hold the spoke firmly. The second method permits dismantling and re-assembly of the model. The kit provides all necessary equipment including a stand, supports for the base pairs, glue and an instruction booklet. The spokes are in four lengths identified by their colours. These correspond to single bonds, double bonds, bonds from hydrogen to a different element (the shortest) and hydrogen bonds (the longest). The assembly of a DNA model from this kit is currently a student project at Westfield College.

Kits designed for the assembly of proteins and nucleic acids, including DNA and RNA, can be purchased in the Atomunits system. A part-assembled Minit DNA can be obtained as well, but complete assemblies are not common among manufactured skeletal models, presumably because assembly is usually easy with these systems. Anderson [4] reports the construction of a DNA model from FMM tubing attached to a vertical aluminium support bearing narrower gauge lateral branches to which the base pairs and sugar phosphate groups were attached at appropriate intervals and rotations. Most of the model was assembled by students as part of a project. The model is similar to that of Watson and Crick [5] except that plastic tubing is used instead of brass rod. The original Watson and Crick model of DNA, which is of the Kendrew type, is a powerful demonstration of the inspiration which a model may provide in studies of the structures of molecules and crystals. Watson's statement, 'The next scientific step was to compare seriously the experimental X-ray data with the diffraction pattern predicted by our model' and the acceptance by a co-worker that, 'our hooting about model building represented a serious approach to science' confirms my belief in the utility of model building. Moreover, when Watson also says, 'the structure was too pretty not to be true', I am equally convinced of the significant part played by aesthetic judgements in the study of chemistry. However it may be safer to base one's conclusions on more objective criteria!

Another example of the value of this type of model in research is given by Kim and his co-workers [6] who used a model of yeast phenylalanine transfer RNA to clarify the three-dimensional tertiary structure of this substance. The positions of all the nucleotide residues were defined as a result of this work. A Kendrew-type model of myoglobin was used by Keefe and Howe [7] in the teaching of science to medical students. It was constructed of brass rod, with a scale of 20 mm $\equiv$ 0.1 nm, and mounted on a 'tree-style' plastic support.

The Nicholson Molecular Models were designed exclusively for the construction of macromolecules and so they are not described elsewhere in this book. In my opinion they are skeletal models, although some atoms are represented by spheres, the units being joined by pushing a rod into a tube (see Dreiding and Fieser models, Chapter 5). The moulded plastic units may represent either one atom or a group of atoms such as guanidine or indole. The torsional and dihedral angles can be determined by means of a keyed protractor supplied with the kit. Rotational friction at the link is high so that spatial orientation is retained during assembly.

Kits for the construction of a range of enzyme structures and a number of haem proteins are available. With many protein model kits a list of dihedral angles, support rod positions on the base-board and other data to facilitate assembly are supplied. Although individual components are cheap, inevitably the total cost of the largest models such as that of carboxypeptidase A and the complete haemoglobin model (with all four haem groups) is high.

A different colour code is used from the conventional one in order to differentiate, not atoms, but structural entities in the model. The peptide backbone is white, side chain carbons grey, aromatic rings purple, amino and guanidinium groups blue, carboxyl and hydroxyl red and sulphydryl yellow. The scale is 10 mm ≡ 0.1 nm.

The units may be readily modified if required by cutting, drilling or pinning to remove or add parts. Permanent distortion may also be brought about if they are bent after warming and then cooled.

No hydrogen bond models are supplied but wire may be used to simulate H-bonding because it is easily pushed into the units. The wire may be covered with red plastic sleeving if required. In nucleic acid models, stainless steel rod is preferable to wire because it is stronger. Alternatively, magnets may be used in order to illustrate the donor-acceptor relationship of base pairs in DNA.

In addition to proteins and nucleic acids it is possible to construct carbohydrate models with the Nicholson components.

### 6.2.2 Specific Designs for Macromolecules

In these models the principle of representing each atom and/or bond individually is abandoned and the model units represent a group of atoms which may be part of a backbone or attached to it. There are both commercial and homemade models in this category.

#### 6.2.2.i Biobits [8], [9]. [10], [11] (Figure 6.1)

Invented by Dr. Ivor Smith of the Courtauld Institute of Biochemistry, Biobits are designed for the construction of models of nucleic acids and proteins. Instead of the atom-by-atom construction principle a different code is used in which model units represent macromolecular backbones, bases or R groups of amino acids.

The nucleic-acid backbones consist of either repeating units of 5′-phosphodeoxyribose (DNA) or 5′-phophoribose (RNA). In the Biobits system the repeat unit consists of a 20 mm long piece of PVC tubing (green for DNA and yellow for RNA) at one end of which is a purple band representing the phosphate group. Near the band, holes are punched for the reception of ‘bits’ representing purine or pyrimidine bases. These 'bits' are of the 'poppet-bead' type in various shapes and colours with a peg at one end and a hole at the other into which a second 'bit' can be inserted to indicate substituents. For example, a cytosine 'bit' can be converted to a 5′-methyl cytosine unit in this way. Plastic coated wire representing hydrogen bonds is also inserted into the holes. Red-coated wire corresponds to the adenine/thymine (double bond) base pair and blue-coated wire to the guanine/cytosine (triple bond) base pair. As a result it is possible to join two strands to make a double helix and then to illustrate both the replication process and also transcription by the addition of a strand of ribonucleotide. The 20 mm monomeric units are joined by pieces of nylon tube which fit firmly into the PVC tubing. A double-stranded DNA model built in this way and showing over two turns of the helix weighs less than 200 g and is less than 400 mm high. It is therefore easy to store, move around and examine. Chemical reactions such as the substitution of one base by another or enzymic hydrolysis may easily be demonstrated, by changing the 'bit' or breaking the chain in the appropriate places respectively. It is also possible to demonstrate the various forms which may be assumed by a molecule such as those of alanine *t* RNA [8]. Four nucleic acid kits are available, one of DNA/RNA for illustrating replication and transcription and three specific RNA.

The protein Biobit models are constructed according to the same principles, the backbone being blue with red bands indicating the amino and carboxyl groups of the amino acids respectively. The common R groups and other residues such as sulphate, acetyl, formyl and so on are each represented by a specific 'bit' which is attached to the backbone as described above. Modifications of R groups are also accomplished in the same way. That is, an extra 'bit' is added to the one already attached, an OH 'bit' added to a proline 'bit' giving hydroxyproline, for example. Sulphur bridges and hydrogen bonds are indicated by coated wire connectors.

Various auxiliary 'bits' representing groups such as dinitrophenyl (DNP) or dansyl (DNS), acyl and monosaccharide among others, enable analytical and synthetic processes to be illustrated. Chemical attack, for example by enzymes, is indicated by splitting the chain into units representing the known products. For permanent models a continuous piece of tubing is used for the backbone, stiffened by the insertion of plastic-coated wire which protrudes at either end. Each protruding section is bent back and pushed into the tube so that it is firmly held. The tubing may be purchased with the red sleeves already positioned at 20 mm intervals and holes for reception of the bits punched. The tube is then bent according to the known backbone structure, helical portions being formed by winding round a piece of glass tubing about 20 mm in diameter and adjusting the pitch to give the correct arrangement of residues for each complete turn. The R group 'bits' are then inserted into the holes so that they project inwards or outwards as appropriate. It can be seen, for example, that in the case of haemoglobin all the nonpolar groups are internal while the polar groups are on the external surface. This demonstrates nicely why the molecule is so soluble in water.

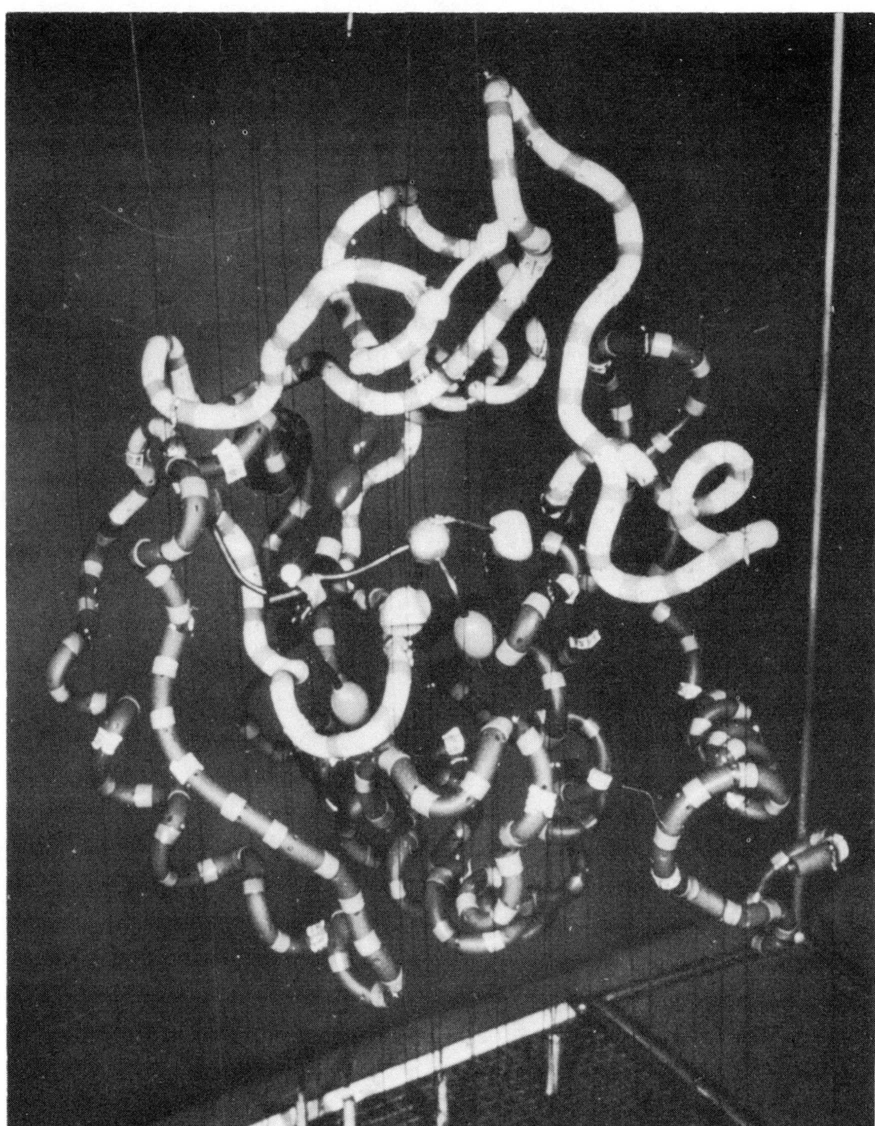

Fig. 6.1 – Biobits: carboxypeptidase A.
*(Photograph supplied by Dr. Ivor Smith, Courtauld Institute of Biochemistry, Middlesex Hospital Medical School.)*

The special advantage of Biobits is that tertiary structure is revealed with great clarity. It is particularly satisfying to be able to see at first glance clefts in the structure into which other groups may fit. For example, in the case of myoglobin the haem cleft is clearly visible and the location of the haem group obvious even if it is not present. A sheet of red, transparent Lucite is used to represent this group. In the case of enzymes, the cleft indicates the site of enzyme-substrate interaction. Biobits were designed specifically as learning aids and students can learn a great deal while constructing them.

A support frame for permanent display models is available. This is of metal and consists of two horizontal pieces of mesh at the top and bottom with four vertical rods joining each of the corners, Spring-loaded wires are arranged vertically, each with wire loops passing through a ferrule which can move up and down. These loops are used to support the backbone and once it is in position the ferrules may be pinched with the crimping tool included in the kit to prevent any further movement. If this fixes the Z co-ordinate, then arrangement of the spring-loaded wires on the two meshworks (mesh 9.6 mm) determines the XY co-ordinates. The scale is 30 mm ≡ 0.2 nm for both nucleic acid and protein Biobit models.

There is a general protein kit, otherwise the kits are for the construction of specific proteins including lysozyme, myoglobin, chymotrypsin, insulin and haemoglobin (complete tetramer). Pre-assembled models may be purchased as well as separate components.

#### 6.2.2.ii Other Systems

Of the commercially-available models, the one developed by Bennett [12] for the study of protein synthesis is worthy of note even though it is not a structural model in the same sense as those described above. The components are of cardboard and consist of a cottage-loaf shape representing a ribosome, four strips each with a sequence of bases marked on them denoting *m* RNAs, *t* RNA segments each marked with an anticodon of three bases and each with a corresponding amino acid unit associated with it. The ribosome unit can be arranged so that *m* RNA strips can be passed through the centre exposing the three-base codons in sequence. Each of these is then matched by interlocking with the appropriate anticodon and subsequently with the corresponding amino acid. The amino acid pieces can be linked in sequence and the protein built up. A helpful and instructive booklet on the ‘Elements of Protein Synthesis’ goes with the model. Together they provide a useful introduction to the subject.

Apart from indicating groups in the Biobits system, poppet beads can be joined together either to form chains representing polymers or to illustrate the principles determining heredity. They can be obtained from a number of sources. Griffin and George Ltd. market them in three colours but the Biobits types provide many more variants in colour and shape and these can be purchased separately.

In addition to the conventional models described in 6.2.1, SRM produce a so-called ‘Framework DNA Model’ made of hardwood. Flat pieces representing the base pairs are attached to the sugar phosphate backbone and to each other by metal fasteners. The backbone segments are joined by metal rods. The base pairs can be uncoupled to demonstrate the replication process, while prongs and receptors between base pairs simulate hydrogen bonding. The set consists of five each of cytosine, guanine, adenine and thymine units, twenty backbone segments, a support, storage box and instructions. The completed model is over three feet tall and clear enough for demonstration purposes in a large room.

Another SRM model is similar in principle to Bennett's and also used to demonstrate protein synthesis. Squares bearing the initial letters of the four bases are magnetically attracted to a metal tray which itself is attracted to a display stand. Consequently any *m* RNA sequence can be arranged. This is matched with *t* RNA anticodon groups and then with the corresponding amino acids. Each anticodon group of three bases can be attached by a snap fastener to the wooden square representing the appropriate amino acid. The amino acid units can be linked together in the same way and a protein sequence built up. The model is of demonstration size with an adequate number of components. Instructions are provided.

Folded paper versions of a DNA model were available a few years ago (for example, see [13] and Figure 6.2) but I do not know of any currently on the market. The chemical structures are drawn or written on the paper in the appropriate place. However, a homemade model of the α-helix using white card has been described [14] where the bonds are drawn on in ink and coloured discs which are stuck on represent atoms. Plastic tubing is also stuck to the card so that the cards, representing peptide groups, may be joined together by means of small tetrahedra representing α-carbon atoms.

Many homemade devices for modelling macromolecules are described in the literature. Coloured pipe-cleaners may be employed for teaching basic facts about helices [15] or for building enzymes and other proteins [16]. Nine colours are reported to be available and pieces may be joined by simply twisting them together after cutting to size. Vedvick and Coates [17] made a model of the quaternary structure of haemoglobin by bending suitable gauge wire into helical and nonhelical portions and using brass sheeting for the haems. These were soldered in position. Each of the α and β chains were supported by pieces of copper tubing representing the rotation axes. By moving a lever it was possible to rotate each chain to a limited extent. The purpose of this was to simulate the structural changes which are believed to occur when the molecule changes from the oxygenated to the deoxygenated form. It is thought that the α chains rotate 9.38° and the β chains 7.43°. The model is capable of demonstrating these processes. A simple *t* RNA model is described by Dugas [18]. It consists of a backbone of 1.5 mm aluminium rod and some forty plastic plates made from 1.5 mm thick polypropylene sheet. These simulate the Watson-Crick base pairs. A cardboard former is used to twist the rod into a helix. Incisions at either end of the plates enable them to be snapped on to the rod. Parti-

Fig. 6.2 – Paper model of DNA.
*(Photograph supplied by Mr. F. Whetstone, Chemistry Department, University of Nottingham).*

cular plates can either be identified by press-on letter sets or coloured along the edge with a marker. The scale is approximately 10 mm ≡ 0.35 nm.

Two membrane models reflect the current interest in this topic. One simulates a lipid membrane with adsorbed chlorophyll and cyanine dye. This is made from polystyrene balls, to represent the hydrophillic head, and pipe-cleaners, representing the hydrophobic tail [19]. The polar heads (spheres) are outermost. Adsorption can be demonstrated by insertion of a square of green pipecleaner (chlorophyll) and red pipecleaner (cyanine dye). The model of the purple membrane devised by Henderson and Unwin [20] shows a single protein molecule consists of seven α-helices. The helices in the model are formed from stacked plates and resemble some of the models which were made when the structures of myoglobin and haemoglobin were being determined.

Finally, two rather special representations of models of macromolecules found alongside the printed page may be of interest. In one case it would not be true to refer to this as a two-dimensional representation [21] because the pictures of lysozyme and its hexasaccharide substrate are made by the Xograph process. This achieves a three-dimensional effect by dividing the picture into many parallel vertical strips and coating them in such a way that parts of the image are diminished or enhanced in the viewer's eyes and a three-dimensional effect results. The second example can be found in the illustrations to the article by Phillips [22] entitled 'The three-dimensional structure of an enzyme molecule'. In fact, these are essentially paintings of skeletal models of lysozyme and its substrate. They are very attractive and show as clearly as possible in two dimensions how the substrate fits into the cleft in the molecule.

#### 6.2.6.iii Synthetic Polymer Models

Many of the systems already discussed are either suitable for, or can be adapted to, the construction of polymer models. Unlike naturally occurring macromolecules, where the repeat units are often large and complex, those of the common synthetic polymers are usually small and relatively simple. Therefore the chief requirement is not to have a wide range of *different* units but to have *enough* of each of the few that are necessary. For this reason it is rather extravagant to use the most sophisticated space-filling or skeletal models, unless great precision is necessary or the models are already available. The cheaper ball-and-spoke models are generally satisfactory and capable of demonstrating the salient structural features. I tend to favour ball-and-spring systems for this purpose, particularly if the structure is helical, because it can then be wound round some such support as a retort stand and held in place with adhesive tape (Figure 6.3).

Poppet beads can again be used to demonstrate the properties of chain polymers or, alternatively, polystyrene or cork spheres can be threaded on string [23] or elastic [24]. The advantage of the spheres is that extra holes can be made in them to enable the chains to be cross-linked. Wire can be used for

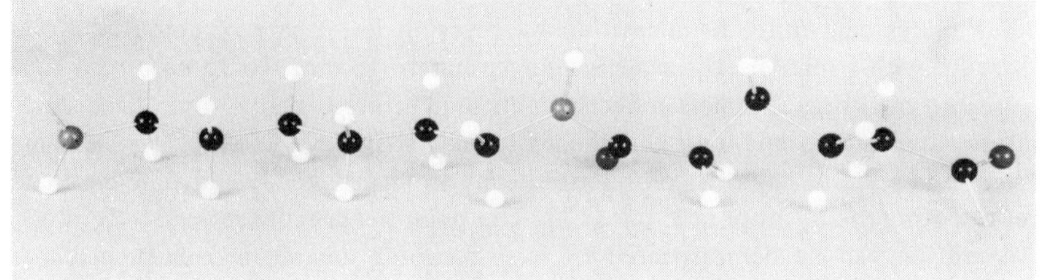

Fig. 6.3 – Gallenkamp Linnell: ball-and-spring model of the repeat unit of Nylon-6,6.

this purpose. Kaye [25] has used plastic coated clothes-line wire with pipe cleaners as side-chains for the construction of models of vinyl isotactic and syndiotactic polymers. After cutting the wire to form the backbone the cut ends must be crimped because these are sharp and could be dangerous. The repeat distances are marked on the backbone and the wire bent at these points so as to give the tetrahedral angle. Pipe-cleaner 'side-chains' are then attached.

A variant on the paper-folding technique is described by Rodriguez [13] who used transparent sheets of cellulose acetate, or similar material, folded into appropriate shapes. Because many polymer structures are helical it is possible to classify them according to the number of monomer units for a given number of complete turns. For example, an $H3_1$ helix has three monomer units in one complete turn. A model of this can be made by making a triangular prism out of the transparent plastic sheet and then drawing the carbon-carbon bonds on the appropriate edges and faces while maintaining the tetrahedral angle. Felt pen, crayon or acetate-based drawing inks may be used. An alternative, faster method is to make an overhead projector transparency of a drawing using a suitable copying machine. For helices with an even number of monomers for each repeat unit a square mandrel should be used. Pendant groups are drawn on separate pieces of sheet and then inserted into slits in the backbone mandrel at suitable intervals and angles. In this way both isotactic and syndiotactic polymers can be simulated. Another method is to draw the complete molecule, including pendant groups, on the sheet and then to cut and fold it so that the correct geometry is achieved. Colour can be added to any of these models using felt-tip pens or crayons. Rodriguez has also described a model of a polymer molecule where the main carbon atom chain is represented by aluminium label tape, crimped at appropriate intervals so that relative orientations are correct. Details of the heavy metal jig used for crimping are given [46].

## 6.3 EXTENDED ARRAYS AND POLYHEDRA

This Section is concerned solely with inorganic structures including those large assemblies which may be regarded as the analogues of the organic macromolecules. Although these large assemblies can be built up from conventional ball-and-spoke or skeletal models, many of them can be described in terms of linked polyhedra and so it is often both convenient and instructive to construct them in this way. Discrete polyhedra are first discussed and then the ways in which some of these may be linked to form structures which are infinite in one, two or three dimensions. Models of the crystal systems and related species are also considered.

### 6.3.1 Conventional Models

It is obvious that no molecular space-filling models are suitable for ionic lattices, while space-filling ionic models are inappropriate because they are constructed from spheres (Chapter 3). Of the ball-and-spoke models, permanent ones such as those of extended arrays produced by CSL (Chapter 4.2.2) are good but both expensive and large, which poses the problem of storage and display. Hence it follows that if the user wishes to construct these large assemblies of ball-and-spoke models himself the best choices are those types which are small, light and can be built on a small scale, and which are inexpensive. The most suitable on all counts are the Beevers' Miniature Models (Chapter 4.2.1). The scale is 10 mm ≡ 0.1 nm so that the largest of the preassembled models has a volume of approximately only one cubic foot. This is of the rock salt structure with twelve balls on each edge and consists of 1331 small cubes requiring 1728 balls and 4752 spokes for their construction. Another ball-and-spoke kit which could be used for large assemblies is the HGS/Maruzen crystal structure set (Chapter 4.4.9) which contains 292 polyhedra and 588 spokes. The components are of light plastics material and the polyhedra are small. However, the scale is the same as the CSL range at 25 mm ≡ 0.1 nm, which means that the space occupied by an extended array would be considerable. Of course, single unit cells of some of the common crystal structures are easily constructed with these models.

Of the conventional models, the skeletal variety is undoubtedly the most appropriate for the construction of large assemblies, especially when the scale is small or can be adjusted to the user's own requirements. Minit models (Chapter 5.2.10) have a small scale, 20 mm ≡ 0.1 nm, adjustable down to 12.5 mm ≡ 0.1 nm, and the atom units have the right numbers of prongs and appropriate geometries for inorganic structures. The related Orbit system also has suitable components but they are larger. The manufacturer's recommended scale is 30 mm ≡ 0.1 nm, though this could be somewhat reduced. Special kits for lattices are available in both these systems.

Because the 'valence clusters' of the FMM system are four-, five- or six-pronged with the prongs pointing towards the corners of a tetrahedron, trigonal bipyramid or octahedron respectively, the structures which can be built are fewer than with the Minit and Orbit systems. However, a very large number of structures are formed from linked tetrahedra or octahedra and for these the FMM components are highly satisfactory. I have made up anionic frameworks of the synthetic zeolites using FMM black tubing cut into 30 mm lengths (the tubing is sold in 120 mm lengths) and the tetrahedral valence clusters [26]. The Si-Si(A1) distance is 0.309 nm [27] so that, allowing for the width of the valence clusters, a reasonable approximation to a scale of 10 mm ≡ 0.1 nm is achieved. An additional advantage is that mechanical constraints cause the tubing to bend slightly making it correspond roughly to the 140° Si-O-Si(A1) angle which exists in the real framework. The tubing may be clad with one inch diameter polystyrene spheres which have been cut in half and a groove gouged out for the tubing. They can be fitted over the tubing and glued together again. Ideally the spheres should be 27.0 mm instead of 24.5 mm in diameter, but these are the nearest obtainable. Models of possible sorbate molecules can then be made from molecular space-filling models with the same scale and it can easily be shown whether or not these are small enough in cross-section to pass through the 'windows' into the channels. FMM components also can be used to show how tetrahedra and octahedra are linked by corners or edges using linear or angular fasteners or one of the 'clusters' to form infinite arrays such as the layer lattice of chromium (III) chloride and polyions like the heptamolybdate ion (Figure 6.4).

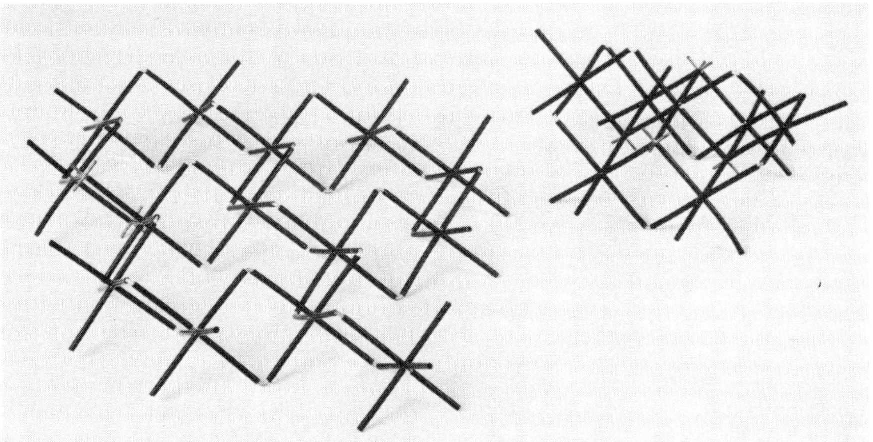

Fig. 6.4 – FMM: (a) layer lattice of chromium(III) chloride, (b) the heptamolybdate anion ($Mo_7O_{24}^{6-}$).

Homemade skeletal models of polyhedral borane and metallocene structures have been made by Lam, Lau, Lau and Mak [28] using wooden cocktail sticks and swab-sticks (Q-tips). These can be sharpened either end and used for 'slant carbon-to-metal bonds' as in uranocene, where the uranium atom is represented by a sphere of modelling clay.

Three of the above authors have also recently described some elegant arrays of polyhedra constructed from plastic drinking straws [47].

### 6.3.2 Discrete Polyhedra

Of course, these may be made from the skeletal models just mentioned and also from Geodestix, Geoframe or similar components. Various sets are commercially available. These include a set of miniature geometric shapes in solid orange plastic from Edmund Scientific, none of which has any dimension greater than 0.75 in. As well as the five regular polyhedra there are various prisms, cones and spheroids, some of which would be of little use to chemists but which would be useful mathematical aids. Polyhedra measuring up to nearly five inches in height and made of transparent plastic are also available in a set from Edmund Scientific. Again, the essential ones are included but there are also others which a chemist would not need. They are attractive but expensive. A cheap alternative is to purchase the twelve cutouts from Edmund. These are on card and can be cut, folded and glued to make the polyhedra.†

Until recently a similar kit was available from SRM, and it may become so again if the demand warrants it. This is the Polyhedron kit, which provides cardboard cutouts in various colours and instructions for producing twelve shapes. These comprise the five regular solids, four prisms (including hexagonal and trigonal) and square, trigonal and hexagonal pyramids. The shapes can be used to make a 'Polymobile' (figure 6.5), which is fun to make and a good talking point. The instructions provided are simple and clear and a bright twelve-year-old could probably make the 'Polymobile' unassisted. Other useful items from SRM are the five regular solids in expanded polystyrne, tetrahedron, octahedron, cube (quaintly called hexahedron), icosahedron and dodecahedron. These may be purchased either as a set of five, which is useful for symmetry discussions in

Fig. 6.5 – SRM: 'polymobile'.

†Unfortunately the polyhedra and cutouts do not appear in the latest Edmund catalogue.

small groups, or in fifty or one hundred packs all of the same kind. Also available are cubo-octahedra, rhombic dodecahedra and rhombohedra. Edge lengths vary from 15 to 51 mm, the larger sizes being more common. The Wells' models kit from Oxford University Press includes hollow tetrahedra and octahedra of transparent plastic (Figure 6.6) which may be purchased separately.

Fig. 6.6 – Wells: rutile ($TiO_2$).

The crystal systems may be obtained in transparent plastic ready-made from LaPine Scientific Co. Axes are indicated by coloured strings with equal axes the same colour. These models are large, the longest diagonal being six inches. Also from LaPine is a relatively inexpensive crystal systems kit for assembly by the user. The material used is again transparent plastic and axes can be indicated by elastic cord positioned during assembly. The tetragonal, orthorhombic, hexagonal monoclinic and triclinic unit cells are 3.5 in. tall while the cube is 2.75 in. The models can be adapted to demonstrate the Bravais lattices and illustrated instructions are supplied.

Demonstration-size coloured cardboard models of some common crystals including sulphur, copper sulphate, zircon, quartz and alum and also of the Bravais lattices can be made from two kits available from SRM. Models up to 175 mm high are made from the crystals kit while the longest axis produced from the Bravais lattices kit is 190 mm. Wooden models of the six crystal systems as a set can also be obtained from SRM. These are 180 mm high with Miller indices and relevant axes printed on each face. Another model from SRM which would be useful used in conjunction with those just described is the crystal plane model which is made of strong cardboard. The planes are marked on the model and axes are indicated by metal rods. The model is supported on a stand and together these reach a height of 300 mm.

A good range of polyhedra is available from Polyhedral Solids either in several different colours or uncoloured in expanded polystyrene. From the variety of shapes supplied, a number of additional polyhedra can be assembled. At the corner of each polyhedron is a small depression into which could be glued a sphere, representing an atom or ion. The polyhedra may be glued together and used to represent many well-known inorganic crystal structures.

Where several sets of models are required for class use, prototypes may be made and then duplicated. A simple technique is described by Salmon and Polley [29], who made moulage moulds of their prototypes and used these to produce plastics casts of all the crystal symmetry classes. The moulage used was inexpensive, consisting of agar, glycerine, magnesium sulphate, methyl cellulose and water. Various 'liquid plastics' may be used for casting. For example, Griffin and George market a suitable polyester resin, hardener and colouring paste.

Much time can be saved by purchasing paper or cardboard cutouts of polyhedra, but the patterns and techniques can be found in the literature for those who wish to make their own. Paper tetrahedra have certainly been in use for at least one hundred years. In 1875 van't Hoff sent some which he had made himself to a friend and these were used to demonstrate the principle of the asymmetric carbon atom [30]. In 1968 Yamana reported [31] a technique for making paper tetrahedra from envelopes and this idea was extended by Freeland and O'Brien [32] who used envelopes to construct models of octahedra, pentagonal and trigonal bipyramids and square pyramids. A variant on the regular tetrahedron is the so-called 'Cooper structure' in which all the faces are recessed to simulate the geometry of $sp^3$ orbitals. A model of this structure made from paper has been described by Walker [33].

An easy method for making paper octahedra is given by Ganesan and Renganathan [34], while the chirality of octahedral tris-chelate complexes is demonstrated by Alexander [35] using cardboard octahedra made from cutouts. Coloured tape is used to mark the edges associated with the chelate rings on a pair of octahedra so as to give them a different chirality. Each is then placed inside a transparent cylinder (which may be made from plastic sheets used for overhead transparencies) on which coloured tape has previously been placed to mark right-or left-handed helices. When the chirality of the octahedron and helix are matched the apices to which chelate rings are attached are seen to follow the course of the helix on the outside of the cylinder.

A book on crystal structures by Cracknell [36] gives a number of cutout patterns, including the tetrahedron, octahedron, cube, icosahedron, rhombic dodecahedron and truncated cube (cubo-octahedron). Boron hydride structures modelled from cutouts in paper or card are described by Greenwood and Smith

[37]. Among these are to be found the icosahedron, icosahedral fragments, the pentagonal bipyramid and the $B_{10}H_{10}^{2-}$ ion. Larson has demonstrated the ways in which paper strips which he calls 'Cubelts' can be folded repeatedly to form a variety, not only of polyhedra, but also of molecular structures [38]. A kit of items can be obtained from him at low cost.

Paper, or even card, models are inevitably rather fragile and unable to withstand frequent handling. Expanded polystyrene is more robust, but solid, hard plastics or wood are the most durable materials. Detailed instructions for the preparation of wooden tetrahedra and octahedra are given by Sheppard [39]. These are very easy to follow and it would be simple to use them to make tetrahedra and octahedra out of expanded polystyrene as well, using a hot wire cutter (see Chapter 9) instead of a saw. A simple technique for making tetrahedra and octahedra from rigid polyurethane is described by Mrvosh and Daugherty [40]. The material can be purchased in sheets from which triangular prism shapes are cut with a motorised saw. These can then be sectioned to give alternating tetrahedra and octahedra. The polyhedra were cemented together to form structures of heterpoly anions and silicates.

Olsen and Tobiason [41] describe how a perspex cube designed to hold photographs can instead hold fractions of cork or polystyrene spheres in appropriate places to indicate the occupancy of various sites. For example, for the sodium chloride structure each corner would hold 1/8 of a sphere and each face centre 1/2 of one, representing the chloride ions. The body centre would contain a whole sphere and the edge centres 1/4 spheres representing the sodium ions.

### 6.3.3 Linked Polyhedra

Many inorganic structures can be described in terms of linked polyhedra of various kinds, which Wells has called the topological approach [42]. However, for practical model-building purposes we are virtually confined to a consideration of linked tetrahedra and octahedra. Because of the lack of availability of the building units†, chains of trigonal dodecahedra such as occur in the $ZrF_6^{2-}$ ion, the filling of space by truncated octahedra and various structures formed from tricapped trigonal prisms (for example, $PbCl_2$ and $UCl_3$), among others, are not so easy to portray in this way. The less regular polyhedra are not usually commercially available so it is almost always necessary to make them from paper or card cutouts and this can be a very time-consuming and tedious business. I have found that for the less symmetrical structures, skeletal models such as Geodestix (Chapter 5.2.5) are better suited.

For those readers who may find the topological approach somewhat esoteric, it should perhaps be reiterated that no representation, whether a model or a two-dimensional graphical picture, can do more than emphasise for us some aspect of structural reality and that the best compromise can be effected by

†But see Polyhedral Solids (above and appendices).

using as many different representations as possible. Therefore in teaching structural inorganic chemistry, one would like ideally to be able to produce two or three different models of the same structure, at least to indicate that they are all approximations. However, I must admit to some prejudices in favour of the linked polyhedra concept, if only because it is then possible to systematise, classify and present in a logical sequence much of the mass of structural data. Table 6.2 summarises some of the ways in which tetrahedra and octahedra may be linked. Further model-building information and theoretical background can be found in three books by Wells, [42], [43], [44] whose model kit includes hollow clear plastic tetrahedra and octahedra which may be linked by rubber

**Table 6.2**

Linked tetrahedra and octahedra

| Mode of linkage | Structure | Example |
|---|---|---|
| Tetrahedra sharing opposite corners | chain | pyroxenes or $SO_{3(c)}$ ('asbestos' form) |
| Tetrahedra sharing three corners | layer | $P_2O_5$ layer form |
| Tetrahedra sharing four corners | layer | $HgI_2$ red form |
| Tetrahedra sharing four corners | 3-D | ZnS (blende and wurtzite) |
| Tetrahedra sharing two *and* three corners | double chain | amphiboles |
| Tetrahedra sharing opposite edges | chain | $BeCl_2$ or $SiS_2$ |
| Tetrahedra sharing four edges | layer | PbO (layer form) |
| Octahedra sharing opposite corners | chain | $AlF_6$ units in $Tl_2AlF_5$ |
| Octahedra sharing four corners | layer | $AlF_6$ units in $TlAlF_4$ |
| Octahedra sharing opposite edges | chain | $SnCl_6$ units in $K_2SnCl_4.H_2O$ or $HgCl_6$ units in $K_2HgCl_4.H_2O$ |
| Octahedra sharing three edges | layer | $CrCl_3$ |
| Octahedra sharing six edges | layer | $CdI_2$ |
| Octahedra sharing six corners | 3-D | $ReO_3$ |
| Octahedra sharing two edges and corners | 3-D | $TiO_2$ (rutile structure) |
| Octahedra sharing one face, three edges and all corners | 3-D | $Al_2O_3$ (corundum structure) |
| A pair of octahedra sharing one face | dimer | $WCl_6$ units in $K_2W_2Cl_9$ |
| Octahedra sharing opposite faces | chain | $ZrI_3$ |

connectors (Figure 6.6). These components can be purchased separately. (The kit also includes spheres, ball-and-spoke and skeletal components, see Chapters 3, 4 and 5). Expanded polystyrene tetrahedra and octahedra can be obtained from SRM. Alternatively, the sources and techniques given in 6.3.2 (above) may be used to acquire polyhedra which can then be glued, pinned or taped together.

While Table 6.2 does not exhaust the possibilities, it does give some idea of the variety of structure which may be regarded as consisting of linked polyhedra. Many more examples can be drawn from the oxides, halides, silicates and other naturally-occurring mineral structures. There is also a group of dimers typified by $Al_2Br_6$ (two tetrahedra sharing an edge) and $Nb_2Cl_{10}$ (two octahedra sharing an edge) as well as various cyclic compounds which are not tabulated.

Polyions can be formed by the condensation of discrete tetrahedral oxy anions such as phosphate and vanadate which link together to form chains. The well-known isopolyacids of molybdenum and tungsten consist of linked $MoO_6$ or $WO_6$ octahedra, although in this case the monomeric units do not exist. These structures are difficult to explain verbally or by drawings but the models are easily assembled and prove to be very helpful aids to understanding. It is also possible to construct models of heteropolyacids (Figure 6.7).

Fig. 6.7 – Linked octahedra: (a) Wells model of hexamolybdotellurate anion ($TeMo_6O_{24}^{6-}$), (b) SRM polystyrene octahedra linked to form a model of the paratungstate anion ($W_{12}O_{46}^{20-}$).

In conclusion, it is perhaps worth noting that models of linked polyhedra have been successfully used as a means of communication in research. For example in their paper on the structure of erionite Staples and Gard [45] published photographs of a model made by linking stiff paper tetrahedra. Again, paper models showing the mode of linkage of truncated octahedra in the sodalite zeolites have been used as an effective means of visualising these structures (Figure 6.8). Lastly, a substructure of octahedra in the three-dimensional arrays of anatase and rutile (both $TiO_2$) can be identified from models in which the octahedra in anatase share equatorial corners to form a layer while those in rutile share edges to form chains [44].

Fig. 6.8 – Stiff paper model of the X- or Y-type anionic framework of synthetic zeolites and faujasite.
*(Photograph supplied by Mr. F. Whetstone, Chemistry Department, University of Nottingham.)*

**Table 6.3**
Macromolecular models and polyhedra

| Model | Price code (Table 1.4) |
|---|---|
| Edmund pregrouped clusters | (a), (b) |
| Nicholson Molecular Models | (a), (b), (c), (d) |
| Biobits | (b), (c) |
| Bennett's cardboard protein synthesis model | (a) |
| Griffin and George poppet beads | (a) |
| SRM Framework DNA model | (c) |
| SRM Protein Synthesis Model | (c) |
| Edmund Miniature Geometric Solids | (a) |
| Edmund Jumbo Geometric Solids | (c) |
| Edmund cutouts | (a) |
| SRM Polyhedron kit | (a) |
| SRM Five Regular Solids | (a) |
| SRM other polyhedra | (a) |
| Wells (OUP) polyhedra | (a) |
| LaPine Crystal Model Set | (b) |
| LaPine Crystal Systems Kit | (a) |
| SRM Models for Common Crystals | (a) |
| SRM Nature's Crystal Set | (a) |
| SRM Six Crystal Systems (wooden models) | (a), (b) |
| SRM Crystal Plane Model | (a) |
| Polyhedral Solids | (a), (b) |
| Cubelts | (a) |

Note. Order of presentation as in text.

## REFERENCES

[1] Gray, W. R., Sandberg, L. B. and Foster, J. A., *Nature,* **246**, 461 (1973).
[2] Landé, S., *J. Chem. Educ.,* **45**, 587 (1968).
[3] Beevers, C. A., *Educ. in Chem.,* **11**, 198 (1974).
[4] Anderson, J. A., *J. Chem. Educ.,* **49**, 329 (1972).
[5] Watson, J. D., *The Double Helix,* Weidenfeld and Nicolson 1968.
[6] Kim, S. H., McPherson, A., Quigley, G. J., Rich, A., Seeman, N. C., Sussman, J. L., Suddath, F. L. and Wang, A. H. J., *Science,* **185**, 435 (1974).
[7] Keefe, W. E. and Howe, D-B., *J. Chem. Educ.,* **50**, 628 (1973).
[8] Smith, I., *J. Biol. Educ.,* **3**, 193 (1969).
[9] Smith, I. and Smith, M. J., *Educ. in Chem.,* **6**, 2 (1969).

[10] Smith, I., Smith, M. J. and Doré, C. F., Chapter 40 *Chromatographic and Electrophoretic Techniques,* Vol. 1, edited I. Smith, Heinemann (1969).
[11] Smith, I., Smith, M. J. and Roberts, L., *J. Chem. Educ.,* **47**, 302 (1970).
[12] Bennett, T. P., *Elements of Protein Synthesis,* Freeman (1969).
[13] Rodriguez, F., *J. Chem. Educ.,* **45**, 507 (1968).
[14] Davies, A. K., *Educ. in Chem.,* **13**, 71 (1976).
[15] Magliulo, A. R., *J. Chem. Educ.,* **49**, 391 (1972).
[16] Nicholson, L., *J. Chem. Educ.,* **46**, 671 (1969).
[17] Vedvick, T. and Coates, M., *J. Chem. Educ.,* **48**, 537 (1971).
[18] Dugas, H., *J. Chem. Educ.,* **54**, 298, (1977).
[19] Huebner, J. S., *J. Chem. Educ.,* **54**, 171, (1977).
[20] Henderson, R. and Unwin, P. N. T., *Nature,* **257**, 28 (1975).
[21] Harte, R. A. and Rupley, J. A., *J. Biol. Chem.,* **243**, 1663 (1968).
[22] Phillips, D. C., *Sci. Amer.,* **215**, 78 (1966).
[23] Carraher, C. E., *J. Chem. Educ.,* **47**, 581 (1970).
[24] Tetlow, K. S., *Educ. in Chem.,* **1**, 7 (1964).
[25] Kaye, H., *J. Chem. Educ.,* **48**, 201 (1971).
[26] Walton, A., *Educ. in Chem.,* **9**, 146 (1972).
[27] Bergerhoff, G., Koyama, H. and Nowacki, W., *Experientia,* **12**, 418 (1956)
[28] Lam, C. N., Lau, O. W., Lau, Y. S. and Mak, T. C. W., *J. Chem. Educ.,* **53**, 740 (1976).
[29] Salmon, J. F., Polley, S. J. and Polley, C. A., *J. Chem. Educ.,* **50**, 726 (1973).
[30] Snelders, H. A. M., *J. Chem. Educ.,* **51**, 2 (1974).
[31] Yamana, S., *J. Chem. Educ.,* **45**, 245 (1968).
[32] Freeland, B. H. and O'Brien, R. J., *J. Chem. Educ.,* **48**, 771 (1971).
[33] Walker, R. A., *J. Chem Educ.,* **50**, 703 (1973).
[34] Ganesan, L. R. and Renganathan, S., *J. Chem. Educ.,* **48**, 59 (1971).
[35] Alexander, M. D., *J. Chem. Educ.,* **50**, 125 (1973).
[36] Cracknell, A. P., *Crystals and Their Structures,* Pergamon (1969).
[37] Greenwood, N. N. and Smith, J., *Educ. in Chem.,* **1**, 25, (1964).
[38] Larson, G. O. Demonstration of 'Cubelts' reported by Rhodes, G. and Daly, J. M., *J. Chem. Educ.,* **54**, 12 (1977).
[39] Sheppard, W. J., *J. Chem. Educ.,* **44**, 683 (1967).
[40] Mrvosh, M. E. and Daugherty, K. E., *J. Chem. Educ.,* **52**, 239 (1975).
[41] Olsen, R. C. and Tobiason, F. L., *J. Chem. Educ.,* **52**, 509 (1975).
[42] Wells, A. F., *Models in Structural Inorganic Chemistry,* Oxford (1970).
[43] Wells, A. F., *The Third Dimension in Chemistry,* Oxford, (1956).
[44] Wells, A. F., *Structural Inorganic Chemistry,* 4th Edn., Oxford (1975).
[45] Staples, L. W. and Gard, J. A., *Min. Mag.* **32**, 261 (1959).
[46] Rodriguez, F., *J. Chem. Educ.,* **53**, 92 (1976).
[47] Mak, T. C. W., Lam, C. N. and Lau, O. W., *J. Chem. Educ.,* **54**, 438 (1977)

# 7

# Models depicting atomic and molecular orbitals

## 7.1 INTRODUCTION

Because many secondary school chemistry curricula now include the study of atomic and molecular orbitals, a need has arisen for suitable models which can effectively demonstrate these concepts. However, in some respects the portrayal of orbitals by material structures is hedged about with even more difficulties then the portrayal of molecules and crystals by these means, and the models used are invariably inadequate.

As far as atomic orbitals are concerned, it is important to be clear about what it is that the models are depicting. This may be either the three-dimensional angular dependence of the wave function, $\psi$, as in the Norbury models discussed in 7.2 below, or the probability, $\psi^2$, of finding an electron within a given volume [1]. In fact the polar plots of the angular dependence of $\psi$ are without physical meaning, but $\psi^2$ incorporates both distance and direction. This permits the construction of a boundary surface which may be supposed to define the orbital to the extent that there is, say, a 95% chance of one or two electrons being within this volume. Even from a model-building point of view this distinction is not mere pedantry, because the *shapes* are different. For example, the $\psi$ plot of a *p*-orbital results in two spheres in contact, while the $\psi^2$ plot produces a dumbbell shape. Most orbital models employ the probability concept, the advantages being both that students generally find this easier to understand and that the orbital shapes represent something which can be accepted as physically 'real'.

Because the boundary surface cannot itself be precisely defined neither can the orbital models, and the shapes used tend to be even more arbitrary than those used in other models. Also, I know of no manufactured set of models of atomic orbitals which takes account of radial distribution functions and indicates the variation in size of the orbitals with principle quantum number. The chief utility of these models is their ability to demonstrate the shapes of orbitals and their spatial distribution. With their aid it is therefore possible to rationalise orbital overlap, ligand or crystal field theory and similar concepts.

Two theories are commonly used to explain covalent bond formation and the shapes of small molecules. These are Molecular Orbital (MO) theory and the

Valence Shell Electron Pair Repulsion (VSEPR) theory. The first incorporates the concept of hybridisation while the second is based upon the mutual repulsions of bond-pairs and lone-pairs. The MO theory can be found in most books on chemical bonding and the VSEPR theory in References [2] to [5], among others. Some of the manufactured kits of orbital models provide suitable shapes for simulating $\sigma$-bonds, $\pi$-bonds, bond pairs and lone-pairs, but if the required components are not included it is often possible to purchase these elsewhere separately, or to fabricate them. Some suggestions and guidelines are set out below.

As always, the chemical education literature contains a number of articles on homemade orbital models, many dating from the period (about ten years ago) when few manufactured models were available. A variety of techniques and materials has been used and applications are diverse, though mainly involving teaching. Apart from non-availability of commercial models, the incentive to devise and construct custom-built, low-cost models has clearly been strong and much ingenuity has been shown. The advent of expanded polystyrene has brought about a minor revolution in model-making and this is particularly apparent when the materials which used to be used for orbital models are considered. These are either less tractable, heavier and less easily or speedily fabricated than expanded polystyrene or more expensive. Examples are papier-mâché, which is time-consuming, cork, which is difficult to shape, and plasticine, which is heavy. Some conventional models (particularly skeletal ones, see Chapter 5) can be adapted by adding attachments of various kinds so that the symmetry of the relevant orbitals is indicated or their approximate geometry outlined. These and other two-dimensional representations are discussed in Section 7.4.

## 7.2 MANUFACTURED ORBITAL MODELS: ASSEMBLED, COMPLETE OR PARTIAL KITS

### 7.2.1 Griffin d-Orbital Electron Model (Figure 7.1)

This model consists of a vertical rod to which four cross-pieces are attached at intervals. The orientation of these with respect to each other is such that they can be used to represent the symmetry axes of the $d_{x^2-y^2}$, $d_{xy}$, $d_{yz}$ and $d_{xz}$ orbitals. On to each of the four arms of each cross-piece a coloured lobe can be slipped, a different colour being used for each orbital. Permanently installed at the top of the vertical rod are two vertical lobes and the encircling middle band representing the $d_{z^2}$ orbital. Spokes radiating from the origin of the $d_{z^2}$ orbital model indicate the symmetry axes of the other d-orbitals. The lobes portraying the $d_{xy}$, $d_{yz}$, $d_{xz}$ and $d_{x^2-y^2}$ orbitals can either be placed on the individual orbital axes or all combined on the uppermost set of axes incorporating $d_{z^2}$. In this way, it is possible either to demonstrate the approximate spherical symmetry when all the orbitals are combined or to show the symmetry of individual orbitals. A small part of each spoke protrudes when the lobe is in place so that a piece of plasticine or a small polystyrene sphere can be attached to indicate a point charge or potential ligand position. Hence the model can be used very effectively in discussions of crystal and ligand field effects. The kit includes a booklet which describes the situation in an octahedral ligand field. No support base is supplied but it can be purchased separately. In fact a retort stand base with a hole of suitable size drilled in it would be perfectly adequate. The fit needs to be good, because when all the lobes are combined the model is rather top-heavy. A similar model is described by Johnstone [6].

Fig. 7.1 – Griffin d-orbital model.

**7.2.2 Norbury Models** (Figure 7.2)

A complete set (*s, p, d* and *f*) of orbital models was designed by the late Dr. A. H. Norbury [7] some years ago and these sets are marketed by CSL in Britain and Klinger in America. The models represent three-dimensional plots of the angular variation of the wave function. Consequently $\psi$ rather than $\psi^2$ is portrayed, which means that lobes can be regarded as positive or negative and can be coloured appropriately (red or yellow) in the models. Such a distinction is of great value when discussing orbital overlap in MO theory or orbital symmetries. However, the designer emphasised that these models do not actually show us what an orbital 'looks like'. What they do show, and accurately, is the magnitude of the angular wave function in any direction, because it is proportional to the distance between the origin and the surface in that direction.

The models are of moulded polystyrene and supported on a stainless steel arrangement of $x$, $y$ and $z$ axes so that the assembly is about ten inches high. The complete set consists of one $s$, three $p$, five d and seven $f$ orbitals, all on supports, together with twenty $p$ orbitals and twenty-two $sp$ hybrid orbitals unsupported, nine 'tetrahedral frames' and sixty small white spheres. The unsupported orbital models can be attached to the axes of the supported models so as to indicate bonding situations and the formation of molecular orbitals. Six of the white spheres attached to the axes indicate the positions of point charges in an octahedral field, while the 'tetrahedral frame' can be arranged around the orbital model so that white spheres attached at the ends indicate point charges in a tetrahedral field. In this way crystal field effects on the degeneracy of the orbitals can be readily demonstrated. Of the hybrid orbitals, thirteen have large positive (red) lobes and nine have large negative (yellow) lobes. These can be used to illustrate metal-ligand orbital overlap in complexes.

Fig. 7.2 – Norbury orbitals: (a) $d_{z^2}$, (b) hybrid, (c) $d_{x^2-y^2}$.

Neither of the two partial sets contains any $f$ orbital models and one only contains three $d$ orbital models (of the three different symmetries). Both contain fewer of the unsupported hybrid and $p$ orbital components, fewer spheres and fewer 'terahedral frames' than the complete set. All components may be purchased separately.

These models are probably the best available but they are expensive. On the other hand, we do know what they are meant to portray, which is unusual for a set of orbital models, and the contours are accurately reproduced. On academic grounds they are warmly recommended.

### 7.2.3 Probability Envelopes of Electron Location (PEEL) Models (Figure 7.3)

As their name implies, these models represent the probability function, $\psi^2$. The models are essentially of molecular orbitals and can be regarded as representing the charge cloud contours associated with chemical bonds. There are four principal components in the PEEL kits. First the so-called 'protonated' orbital represents the MO associated with a covalent bond between hydrogen and a non-metal [8], [9]. Then two lengths of $\sigma$ orbital equivalent to 0.150 nm and 0.135 nm are used for carbon-carbon single and multiple bonds respectively, the latter including aromatic. Thirdly, there are the nonbonding or 'lone-pair' orbitals, which are red and shorter and fatter than the $\sigma$ orbitals, which are white and lastly, $\pi$ which are green and of three types. Straight ones are associated with two nuclei as in ethylene, for example. Bent ones are associated with three nuclei as in the acetate ion. The ring is for use with aromatic species. All these components are of expanded polystyrene.

Fig. 7.3 – PEEL models: (a) water, (b) ethylene.

The nuclei are Linnell balls and connecting spokes are the springs used with the same models. The PEEL models were designed by Ormerod in conjunction with Gallenkamp and so the orbital models are compatible with the ball-and-spoke models produced by the same company (Chapter 4.4.7). The $\sigma$ orbitals all have indentations where they meet the spheres of the framework so that the appropriate contours and the scale of 60 mm $\equiv$ 0.1 nm are retained. The 25 mm or 38 mm springs either pass through or into the polystyrene shapes. Nowadays the holes into which the springs are inserted are lined with moulded polypropylene inserts to prevent enlargement of the holes and to improve the grip.

Earlier models tended to get damaged rather quickly when handled.

There are three kits. the introductory one enables several small molecules to be constructed including ethylene. It comprises short sigma, 'protonated', lone-pair and pi orbitals, balls and springs, a scale card, a stencil (for drawing) and instructions. The benzene-ring set has only enough components to make the benzene model but the 'General Purpose Set' contains a variety of orbital ball-and-spoke components. A good range of molecular models can be assembled with it. All components may be purchased individually.

The models are useful for teaching up to, and including, first year University level and are inexpensive.

### 7.2.4 SASM Orbital Models

An extensive range of orbital models can be obtained from the French company SASM. The principles of construction are similar to those for the PEEL models (7.2.3 above) in that a ball-and-spoke framework supports the orbital species, but the spheres (with suitably orientated holes) are made of metal, the spokes of plastics and the orbitals of painted wood, coloured so as to distinguish the different types. Both atomic and molecular orbital models are available including *s, p, d, sp*, $sp^2$ and $sp^3$ species. Other hybrids and *f* orbital types can be supplied if specially requested. The ususal spheres or oblate spheroid shapes are used for those species just mentioned but two additional series exist. In one the shapes contain extra holes so that they can be linked to form MO models, or alternatively be used to link molecular models as in hydrogen-bonded assemblies, for example. The other series is particularly noteworthy, because concavities on the surfaces of the shapes are made to fit the convex surfaces of other shapes so conveying the principle of orbital overlap. Spokes already incorporated are used to hold the combining shapes together. For example, it is possible to demonstrate orbital overlap of *s* and $sp^3$ hybrid orbitals such as occurs in the water and ammonia molecules by combining an *s* orbital of the first series, with one hole, with an $sp^3$ hybrid lobe of the second series, which has one spoke and one hole. The hole in the latter is linked by a spoke to the nucleus, the metal sphere at the centre of the assembly. There does not appear to be a particular lone-pair model, but the large lobe of the *sp* hybrid shape would serve.

Although it is possible to construct a number of models of small molecules and ions with the $\sigma$ orbitals, $\pi$ bonds are only indicated by plastic rods inserted into the holes in appropriate adjacent orbital lobes. The SASM models are unusual in the variety of species available and in the mechanism for demonstrating orbital overlap.

### 7.2.5 SRM Orbital Models (Figure 7.4)

SRM produce four orbital kits, two of which can be used in conjunction with the same company's demonstration-size ball-and-spoke models (Chapter 4.4.14). The polystyrene orbital attachments consist of red or green 'atomic' orbitals, green 'lone-pairs', red and white 'protonated' orbitals and red $\sigma$ bonds. $\pi$ bonds are represented by flat strips of red polystyrene with holes at appropriate intervals which rest on the tops of orbital lobes. Hence the $\pi$ bonds in ethylene are indicated by two straight, two-holed strips above and below the $\sigma$ bond while bent three-hole strips are used for the carboxyl group and two six-hole rings for benzene. The components are joined by snap-fasteners to the spherical 'nuclei'. This description applies to the set which we have in the Chemistry Department at Westfield College and Figure 7.4 shows these models. However, the latest SRM Catalogue lists components which differ in such minor respects as colour from the kit with which I am familiar.

Fig. 7.4 – SRM demonstration size: ethylene.

A second kit, also of polystyrene and compatible with the above, can be used to demonstrate the *s*, *p* and *d* orbitals and their hybrids. Rods inserted into a central sphere, to which the orbitals are also attached, indicate the *x*, *y* and *z* axes and the whole can be mounted on a stand. Strips representing $\pi$ bonds are included and several models of small molecules can be assembled.

The completed models of the Classroom Atomic-Molecular Orbital Models are 200 mm tall. There are sufficient components for the *s*, *p* and *d* atomic orbitals to be displayed simultaneously. The kit does not contain hybrid orbital models but these can be simulated by combining components. The number of shapes in this set is rather limited and there is no lone-pair species. Rods are used for the symmetry axes and these protrude from the lobes so they make useful points of attachment either for spheres representing ligands or for $\pi$ bond strips. Orbital lobes and $\pi$ bonds are of expanded polystyrene, while spheres, rods and stands or bases are hardwood. Instructions are included.

A special kit which is designed only for use with Minit models (Chapter 5.2.10) also has expanded polystyrene components. These include *s, p* and *d* orbitals and their hybrids, lone-pair orbitals, $\sigma$ and $\pi$ bonds. A *d*-orbital model supported on a stand is very similar to the Griffin model (7.2.1 above). The usual shapes are used, except that the lone-pair orbital is spherical. Flat strips with holes fitting over the orbital models are again used to portray $\pi$ bonds. These have two, three or six holes, being straight, bent or hexagonal as before. The kit includes the Minit components to which the orbital units are attached and also instructions.

## 7.3 ORBITAL SHAPES AND CLADDING TECHNIQUES

### 7.3.1 Use of polystyrene

Although most of the kit manufacturers will sell individual components separately, spheres, partial spheres, egg-shapes and 'teardrop' shapes can also be bought from other sources. For example, Plasteel Corporation produces all these and a spheroid with a concavity which fits a 1.25 inch sphere as well. Again, these are expanded polystyrene. In addition to the four kits, SRM can also supply orbital and sigma bond shapes in polystyrene. All these objects can be coloured as desired and joined together by pipe cleaners or any of the usual methods. Either manufactured ready-drilled spheres can be used at the centre of the assembly to avoid orientation problems or the components may be simply stuck together.

Several writers give constructional details for the preparation of orbital models from polystyrene shapes. For example, Martins [10] made *s* and *p* atomic orbitals and their hybrids and also MOs of small molecules such as water and ethylene, supporting the lobes on a pipe-cleaner framework. Orbital overlap can be indicated by sawing off segments from each orbital component and then glueing them together. While greater mechanical stability of the assembly can be achieved by using frames of welding rod brazed together [11], these are naturally more trouble to prepare. Onyszchuk [12] utilises the 'valence clusters' and tubing of the FMM framework models (Chapter 5.2.4) to support his models of the *s* and *p* orbitals and their hybrids, and also models of the $sp^3d$ and $sp^3d^2$ hybrids. The models produced are just as suitable for teaching the basic tenets of VSEPR theory for molecules having up to six electron pairs in their outer shell. The approach used in the two articles cited is to discuss the two theories and their applications to molecular geometry in parallel.

Bassow describes a simple jig and gives detailed instructions for the preparation of a model depicting the three *p* orbitals and also orbital models of hydrogen fluoride, water and ammonia [13]. The construction of a set of *d* orbitals is described by Wilson, [14] the orbital shapes either being joined by wooden dowels or mounted on brazing rod sections soldered together. Each orbital lobe has a small segment removed from the base so that the space-filling characteristic is more apparent. An 'octahedral field' consisting of a pipecleaner octahedral framework with spheres at all the corners, representing ligands, can be placed around the composite *d*-orbital model, whose lobes can be suitably coloured so as to distinguish $t_{2g}$ from $e_g$ orbitals. A similar tetrahedral arrangement can be used to demonstrate the lifting of degeneracy in a tetrahedral ligand field.

Brown [15] employed polystyrene models of bonding and antibonding MOs to illustrate orbital symmetry effects (Woodward-Hoffman Rules) in Diels-Alder additions and similar reactions. Aluminium tubing was inserted into each of the two lobes representing the hybrid and these were joined by a piece of polypropylene tubing pierced with a hole so that the orbital model could then be threaded on to a wire framework representing the $\sigma$ bonding. An extra advantage was that the hybrid, with its lobes of opposite sign indicated by different colouring, could be rotated about the '$\sigma$-bond' axis and the positions of the lobes reversed.

### 7.3.2 Balloons (Figure 7.5)

The use of modelling balloons to illustrate electron pair repulsions and the shapes of small molecules was put forward by Jones and Bentley in 1961 [16] and discussed further by Jones in 1965 [17]. A pleasing analogy exists between the observed mechanical repulsions of lobes formed by twisting two or more balloons together and electron pair repulsions predicted by VSEPR theory which is based on the Pauli Exclusion Principle. Electrons with the same spin will strongly repel each other. It therefore follows that electron pairs, other things being equal, will keep as far away as possible from each other and will so tend to form symmetrical arrangements when numbers and stoichiometry permit. When two long sausage-shaped modelling balloons are twisted together about their centres and let go, the four lobes, if evenly inflated, will point towards the corner of a regular tetrahedron. Similarly, twisting three balloons together produces an octahedral arrangement. Three lobes, made by twisting one balloon at its centre and then twisting in a 'half balloon', which has been knotted and then cut in two half-way along its length after inflation, produces a trigonal planar arrangement. Twisting a 'half balloon' into a tetrahedral arrangement transforms it into a trigonal bipyramid. Where double bonds are formed, these can be simulated by tying two lobes together with string or an elastic band. So with a tetrahedron, two pairs of lobes joined give the linear shape of carbon dioxide, while one pair joined gives that of carbonyl fluoride. With the octahedron three pairs joined produces trigonal planar sulphur trioxide, two pairs joined, sulphuryl chloride, and so on.

The VSEPR theory states that lone-pair/lone-pair repulsions are greater than lone-pair/bond-pair repulsions which are themselves greater than bond pair/bond-pair repulsions. This effect can be demonstrated if some shorter, fatter modelling balloons are used in conjunction with the long ones. Lobes of the former then simulate lone-pairs and those of the latter bond-pairs so that if one of each kind are twisted together the resulting tetrahedron has two fat lobes which repel one another more than the two thin lobes. This simulates the situation in the water molecule where the bond angle is less than the tetrahedral value. If another thin 'half balloon' is twisted into the 'water' assembly, a rearrangement resulting in a distorted T-shape takes place. This is the analogue of the bonding arrangement in the chlorine trifluoride molecule.

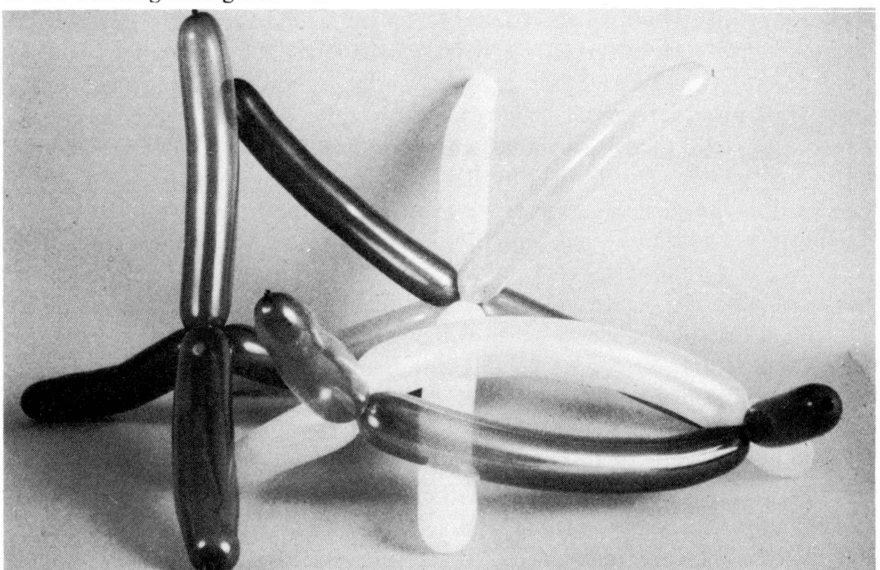

Fig. 7.5 – Modelling balloons: tetrahedron, octahedron and ethylene.

Another use of the modelling balloons is to demonstrate *cis-trans* isomerism in alkenes by taking two long balloons of different colours and twisting them together in two places, near either end. An extra twist through 180° at one end will convert *trans* to *cis* or *vice versa*.

Lastly, $S_N2$ reactions and the Walden inversion can be illustrated by tying in an extra inflated balloon representing the incoming group, and so indicating the transition state. The outgoing group is disposed of with a pin or lighted cigarette and the lobes rearrange themselves appropriately.

More recently, Roberts and Fraynham have also shown how balloons can be used to demonstrate orbitals and molecular geometry [18].

Gillespie [4] suggests a different type of model for illustrating the way in which electron pair repulsions give rise to tetrahedral, trigonal bipyramidal or octahedral symmetries. Pairs of polystyrene spheres are linked by stretching a rubber band between them, either end of the band being inserted into a hole in the sphere where it is held by a toothpick inserted at right angles. Twisting together two pairs gives a tetrahedral arrangement and three an octahedral one, while a group of three twisted with a pair gives rise to trigonal bipyramidal geometry.

### 7.3.3 Other Materials and Techniques Used for Orbital Models

Because of the ease with which irregular shapes can be made, papier-mâché has been used over many years for model making of all kinds. Fowles [19] produced a set of papier-mâché *s* and *p* orbitals, hybrids and MOs of small molecules such as benzene and ethylene. A piano wire or welding rod framework was used. Quite recently (1971), Berry and Hoppé [20] reported a technique for moulding papier-mâché into egg-shapes using a child's toy known as 'nesting-eggs'. This produces several different sizes suitable for making orbital or lone-pair charge cloud models. A model with cork 'orbitals' attached to pegs inserted into a central cube is described by Baker [21]. The pegs coming from the cube face centres indicate the $x$, $y$ and $z$ axes and the positions of the $d_{x^2-y^2}$ and $d_{z^2}$ orbitals, while those emerging from edge centres show the orientation of the $d_{xy}$, $d_{yz}$ and $d_{xz}$ orbitals. Foam rubber or plastic cladding of an FMM framework has been used by Vögtle and Goldschmitt [22] to illustrate orbital symmetry-controlled reactions while Vaughan-Williams [23] describes the use of plasticine for cladding a brass or copper welding wire framework. Different coloured plasticine was used for various orbitals and hybrids.

The use of 'orbital caps' of flexible plastic which can be attached to a 'covalent core' consisting of spheres in tangential contact is suggested by Brumlik [24]. These caps can be used either to represent unshared electron pairs (lone-pairs) or $\pi$ components in a molecular orbital. When used in $\pi$ systems, indentations in these rounded disc shapes imply orbital overlap when they are in contact. Large, hollow, transparent models of AOs and MOs were made by Stone and Siegelman [25] from methyl methacrylate by plastic vacuum-forming techniques using aluminium moulds. Overlap can be indicated by cutting off segments from the two components and then glueing them together. An *s* orbital model is six inches in diameter, a *p* orbital fifteen inches long and a methane model approximately twenty-four inches in cross-section. Although the authors list the ready-made module components of a kit, it does not appear to be commercially available at the moment.

There is also a rather unusual way of visualising orbitals which has an advantage over all the others discussed in being able to portray the *interior* of an orbital and to show the radial distribution of electron density *within* the orbital. This is given by McClellan [26]. Contours of constant probability are calculated from the Schrödinger equation for a given orbital and each value equated to a certain number of dots which could then be marked on a vellum sheet so that dot density is proportional to probability or charge density. Next this pattern is reproduced on six transparent acetate sheets, each of which is then formed into a wedge shape. The six wedge shapes are carefully grouped together and when viewed, preferably in a light-box, a three-dimensional effect is achieved.

## 7.4 SIMPLE FRAMEWORKS AND TWO-DIMENSIONAL REPRESENTATIONS

FMM, Minit and Orbit are among a number of the manufactured framework models described in Chapter 5.2 which can have extra tubing attached so as to outline the shapes of atomic or molecular orbitals or roughly to indicate their geometry. A few ball-and-spoke models such as PMDM or Unit (4.4.12 and 4.4.16) also have this capability. Wire frameworks and cork attachments can be used in conjunction with various conventional models to indicate the

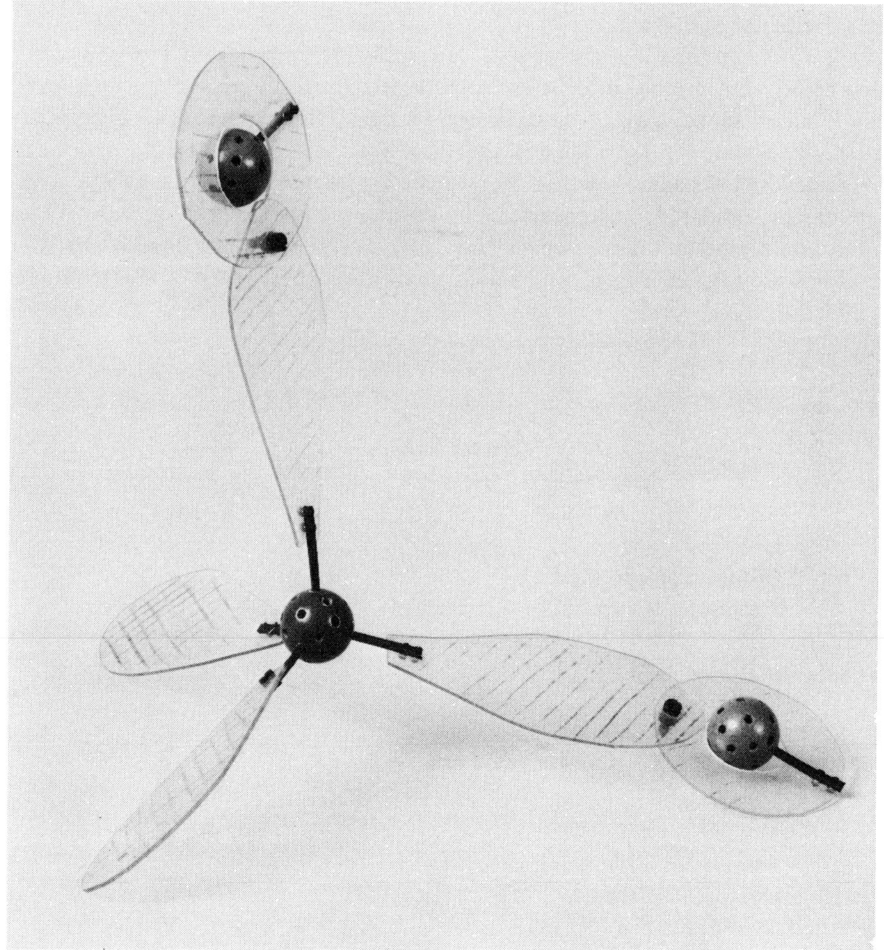

Fig. 7.6 – Perspex sheet used to indicate orbitals in the water molecule.

presence of lone-pair and hybrid orbitals [27]. The FMM 'valence clusters' do not permit *d*-orbital symmetry to be displayed, but a device described by Barrett [28] may be used. This consists of a sphere into which are inserted FMM linear fasteners arranged so that they lie along the *d*-orbital axes. Alternatively, the FMM tubing can be inserted directly into the appropriate holes in a 26-hole sphere (see Chapter 4).

Another homemade model in which only the axes of the *d*-orbitals are depicted is used chiefly to indicate interactions with octahedral or tetrahedral ligand fields [29]. The model consists of a perspex cube with twenty-four inch sides which has at its centre a one inch cube. Projecting from the corners and edge centres are aluminium rods representing the *d*-orbital axes. At the end of each rod is a coloured knob which by its colour and shape assists identification of a particular orbital axis.

Two-dimensional orbital shapes can easily be made from paper or cardboard [15] but a more durable material is perspex sheet [30]. After cutting to the required shape, this can be attached to a CSL slit spoke which has been cut at the slit end so as to provide a flat surface. Both the perspex and the spoke can be drilled and fastened together by means of small nuts and bolts. Orbital overlap may be simulated in a similar way by bolting the two shapes together with an appropriate amount of overlap (Figure 7.6). Perspex is easily marked by wax crayon so suitable markings can be used to indicate whether the orbital is empty, half-full or filled.

**Table 7.1**
Orbital Models

| Model | Price code (Table 1.4) |
|---|---|
| Griffin d-Orbital Electron Model | (b) |
| Norbury models | (b), (c) |
| PEEL models | (a) |
| SASM orbital models | (a), (b), (c) |
| SRM | |
| (1) Orbital Attachments | (b) |
| (2) Giant-size Atomic-Molecular Orbital Models | (b) |
| (3) Classroom Atomic-Molecular Orbital Models | (b) |
| (4) Molecular Orbital Model Kit (for use with Minit framework) | (b) |
| Plasteel polystyrene shapes | (a) |
| SRM polystyrene shapes | (a) |
| Modelling balloons | (a) |

## REFERENCES

[1] See, for example; Greenwood, N. N., *Principles of Atomic Orbitals,* Royal Institute of Chemistry (1964).
[2] Sidgwick, N. V. and Powell, H. M., *Proc. R. Soc.,* **A176**, 153 (1940).
[3] Gillespie, R. J. and Nyholm, R. S., *Q. Rev. Chem. Soc.,* **11**, 339 (1957).
[4] Gillespie, R. J., *Molecular Geometry,* Van Nostrand Reinhold (1972).
[5] Gillespie, R. J., *J. Chem. Educ.,* **51**, 367 (1974).
[6] Johnstone, A. R., *J. Chem. Educ.,* **48**, 74 (1971).
[7] Norbury, A. H., *Educ. in Chem.,* **5**, 28 (1968).
[8] Ormerod, M. B., *School Sci. Rev.,* **49**, (168) 383 (1968).
[9] Ormerod, M. B., (a) *The Architecture and Properties of Matter,* Edward Arnold (1970).
(b) Learning Package; *The Architecture and Properties of Matter,* Scientific Educational Aids, Uxbridge (1976). Includes PEEL models, filmstrip, slides, cassettes, text.
[10] Martins, G., *J. Chem., Educ.,* **41**, 658 (1964).
[11] Lambert, F. L., *J. Chem. Educ.,* **34**, 217 (1957).
[12] Onyszchuk, M., *Can. Chem. Educ.,* April 1968, 17 and October 1968, 5.
[13] Bassow, H., *Construction and Use of Atomic and Molecular Models,* Pergamon (1968).
[14] Wilson, L. R., *J. Chem. Educ.,* **48**, 484 (1971).
[15] Brown, P., *J. Chem. Educ.,* **48**, 535 (1971).
[16] Jones, H. R. and Bentley, R. B., *Proc. Chem Soc.,* 438, 1961.
[17] Jones, H. R., *Educ. in Chem.,* **2**, 25 (1965).
[18] Roberts, R. M. and Traynham, J. G., *J. Chem. Educ.,* **53**, 233 (1976).
[19] Fowles, G. W. A., *J. Chem. Educ.,* **32**, 260 (1955).
[20] Berry, R. W. H. and Hoppé, J. I., *School Sci. Rev.,* **53**, (182), 149 (1971).
[21] Baker, W. L., *J. Chem. Educ.,* **45**, 135 (1968).
[22] Vögtle, B. and Goldschmitt, E., *J. Chem. Educ.,* **51**, 350 (1974).
[23] Vaughan-Williams, H. R., *Educ. in Chem.,* **3**, 156 (1966).
[24] Brumlik, G., *J. Chem. Educ.,* **38**, 502 (1961).
[25] Stone, A. H. and Siegelman, I., *J. Chem. Educ.,* **41**, 395 (1964).
[26] McClellan, A. L., *J. Chem. Educ.,* **47**, 761 (1970).
[27] Kakabadse, G. J. and Theobald, D. W., *Educ. in Chem.,* **4**, 135 (1967).
[28] Barrett, E. J., *J. Chem. Educ.,* **44**, 146 (1967).
[29] Betteridge, D., *J. Chem. Educ.,* **47**, 824 (1970).
[30] Aylett, B. J., personal communication.

# 8

# Dynamic models

## 8.1 INTRODUCTION

Although this book is almost entirely concerned with 'static' models as defined in Chapter 1, a short chapter devoted to dynamic models seems to be appropriate for the sake of completeness. The one feature common to all dynamic models is that some movement of part of the assembly relative to the remainder occurs. Only by using dynamic models can we simulate chemical processes and reactions and so demonstrate various conformational changes. Teachers in particular have clearly felt a need to do this because the chemical education journals contain a number of examples of models which can be used to demonstrate various inversion processes, pseudorotation, $S_N^2$ reactions, racemisation and similar reactions. Another group of models employs magnets as a means of illustrating interactions between atoms and molecules while a third group demonstrates various atomic and molecular phenomena in two-dimensions, often with the aid of an overhead projector.

## 8.2 INVERSIONS AND RELATED PHENOMENA

A series of papers by Berry and his co-workers [1]-[4] describes four dynamic models designed to demonstrate both the nitrogen and Walden inversions and pseudorotation in five co-ordinate compounds. The nitrogen inversion model [1] comprises a tennis ball in which three equidistant slits are cut perpendicular to the equator. These then act as guidelines for the movement of three wooden dowels which are inserted into the slits. At the point where the dowel is held by the slit its diameter is reduced by whittling the wood away evenly. Therefore it cannot slip out easily. At the end away from the tennis ball each dowel has a polystyrene sphere attached to it. Two 20 mm diameter holes are made at the 'poles' of the ball and a small balloon passed through. When this is inflated and the dowels and spheres are at the equator a model of the transition state is produced with two 'orbital lobes' at either end. Application of suitable pressure to a lobe decreases its size so that it becomes negligible compared to the greatly enlarged opposite lobe. The dowels have to be moved manually into the tetrahedral orientation.

A model which works on a similar principle and is also manually operated [2] is used to demonstrate the Walden inversion and similar processes, The model is of the ball-and-spoke type and is commercially available. As it is operated, the incoming group pushes out the outgoing group on the other side of the model of the carbon atom.

The remaining two models are designed to simulate pseudorotation in five co-ordinate compounds. One is easily made from simple materials [3] while the other is more sophisticated in its operation and is commercially available [4]. The chemical principle involved is that certain trigonal bipyramidal compounds, notably those of phosphorus, are able to undergo an intramolecular rearrangement such that the apical and equatorial ligands change places and a new trigonal bipyramid is formed. During the interchange an intermediate square pyramidal structure is formed. Berry has called this process pseudorotation [5]. The homemade model is made from a sponge rubber ball into which is inserted a wooden dowel with a table tennis ball attached at the other end. The rubber ball remains fixed and can be regarded as the pivot. Wedges are cut out of it so that the four other dowels (with attached spheres) can be inserted and moved either in a horizontal or vertical plane (two of each) through 30°. Manual manipulation of the bond angles readily converts a trigonal bipyramidal into a square pyramidal arrangement and further manipulation produces a trigonal bipyramidal structure different from the first in that apical and equatorial species have changed places. In the commercial model the opposing motions required to effect interchange are brought about by a push rod device which operates a system of levers housed in the central aluminium sphere. Consequently all the four interchanging 'ligands' move simultaneously, while in the manual model they have to be moved individually.

Other 'invertible nitrogen' models include a Dreiding type (see Chapter 5.2.2) consisting of two rods and a tube which is pivoted at the angle between the rods by means of a small ring so that it can be rotated [6]. Another, [7] intended to be used as a bridgehead nitrogen atom model in saturated fused rings, has three swivel arms. There are two versions of this model, one of which is compatible with Dreiding models (Chapter 5.2.2) and the other with Courtauld space-filling models (Chapter 2.3.1). The latter is an adaptation of a Courtauld trivalent nitrogen which has a removable cap, fitting either end, which represents the lone-pair. A third type is the 'reaction centre' designed by Peterson [8], with the three swivel arms and a cap which fits either end. This, too, can be used to simulate inversions of atoms in ring systems.

Some new models designed by Professor Dreiding [24] consist of plastic tetrahedra which can be joined by pressing two ends together. However, unlike the stainless steel Dreiding models with rods and tubes, all the connectors are similar. The unique feature of these models is that rotation can occur about the centre of each tetrahedron so that two arms can be rotated with respect to the other two. By this means *cis* isomers can be converted to *trans* and *vice versa,* and because the plastic units are relatively flexible and the connector mechanism is strong *cis/trans* isomerism in alkenes can be demonstrated. Similarly, inversion about the centre of a tetrahedron can convert one enantiomer into the other. Different atoms and groups are represented by different coloured balls attached to the ends of the tetrahedra in appropriate positions. The models can be used to demonstrate stereospecific reactions, orbital symmetry and the Woodward-Hoffman rules.

The models just described were demonstrated by Professor Dreiding at the symposium on Chemical Education in Europe [24] in September, 1975. They have not yet appeared on the market as far as I know. It is possible that the cost proved to be prohibitively high. The Plastic Stereomodels produced by Büchi have the same linkage mechanism (rod into tube) as the stainless steel Dreiding models (Chapter 5.2.2). Rotation about the link is possible, but not at the centre of the tetrahedron. In fact each tetrahedron is supplied in two halves to be glued together by the user. However, if a hole is drilled through the two plastic pieces and a pin passed through instead of glueing, rotation about the centre of the tetrahedron and the operations described above are possible. A plastic pin is preferable as the ends can be flattened after softening by heating.

The modelling balloons described in Chapter 7.3.2 can be used to demonstrate the Walden inversion and $S_N^2$ reactions [9], [10]. A system designed by Nyquist, [11] also for $S_N^2$ reactions, is manually operated like the Berry and Botterill model [2]. Another homemade device for demonstrating Walden inversions and $S_N^2$ reactions [12] consists of three segments cut from a hollow sphere (a toy plastic ball, for example) which together make more than a hemisphere. A rod protrudes from each segment so that the three rods point towards three corners of a tetrahedron. This assembly is then fitted over an inner sphere to which it is attached by rubber bands or tubing. The fourth tetrahedral position is indicated by a rod attached to a cone which fits into the inner sphere, and the opposite pole of the inner sphere can also accept a cone. When inserted, this pushes out the opposite rod and causes the segments and rods of the partial outer sphere to slide over the inner sphere so that an alternative, inverted tetrahedral arrangement is achieved. The device is compatible with the Fieser models described in Chapter 5.2.3.

A pseudorotation model for five co-ordinate compounds, similar in principle to that described above [3], [4] but with a central hard, hollow sphere, has been designed by Riess [13]. The hollow, hard plastic ball is cut through at about a quarter of its diameter and guide grooves cut in the larger piece so that there are two pairs perpendicular to each other. By this means rods bearing the 'ligand' spheres can be moved in the grooves enabling the apical and equatorial ligands to change places. All five rods, including the pivot, are inserted into a rubber stopper which is previously drilled. Reiss points out that drilling is facilitated by first immersing the stopper in liquid nitrogen. With the rubber core and its attachments in place the two parts of the central sphere are rejoined by epoxy adhesive. The core is fixed to the sphere and the pivot by means of nuts because the pivot has a screw thread where it passes through the sphere and into the stopper.

A model for demonstrating racemisation mechanisms in octahedral metal complexes consists of a drilled 50 mm diameter sphere with six holes arranged octahedrally and also holes arranged to coincide with a $C_3$ axis [14]. The sphere is cut in half perpendicular to the $C_3$ axis and rejoined by means of a screw and wingnut so making rotation of the two halves relative to each other possible. Rods are inserted in an octahedral arrangement and chelate rings indicated by joining adjacent pairs of rods with rubber or plastic tubing.

Levenson [15] describes the construction and use of paper models in demonstrating nonrigidity in four- and six-co-ordinate complexes. For example, two 'half octahedra' are made and then joined by means of a rubber band passing through both halves along the $S_6$ axis. Rotation about this axis produces a different octahedral arrangement passing through an intermediate trigonal prismatic arrangement on the way. Such a model may also be used to demonstrate optical activity and racemisation. Interconversion of tetrahedral and square planar complexes may be shown by constructing a four-co-ordinate model in two halves which may be joined and rotated similarly.

## 8.3 USE OF MAGNETS

The so-called 'Self-Crystallising Molecular Models' (Figure 8.1) designed by Kihara [16] are an interesting example of the ways in which magnets can be used to simulate molecular phenomena. Models of molecules which have electrical multipoles are made by incorporating barium ferrite magnets, appropriately polarised in the model. The hexamethylenetetramine molecule, for example, has four nitrogen atoms which together form a tetrahedron. These four negative regions can be simulated in the model by four magnets polarised outwards. The relatively positive regions in between can be represented by four magnets polarised inwards. The eight pieces are combined to form an octupolar sphere. A model of the molecular crystal structure of hexamethylenetetramine can be made by arranging seven octupolar spheres in a layer (six round one) all in the same orientation, placing four more on top, turning the two layers over and adding another set of four to complete the model. The octupolar spheres adjust themselves as necessary to give the correct molecular crystal structure of hexamethylenetetramine, which is body-centred cubic. In this way the interactions between the magnets simulate electrical forces between molecules. It is claimed by Kihara [16] that these forces, when sufficiently large, play a dominant role in determining the structure of the molecular crystal. Models representing carbon dioxide molecules are quadrupolar spheres in which the distribution of the magnets parallels the quadrupole in the molecule. These are easily assembled into the face-centred cubic structure of the carbon dioxide crystal.

Two other model species are available. One is a quadrupolar spindle shape which can be used for monoclinic, orthorhombic or tetragonal structures. The other is an hexadecapolar model consisting of a central nonmagnetic core to which are attached six hemispherical magnets arranged octahedrally. This shape can be used to model hexagonal or rhombohedral structures. The set contains fifteen octupolar spheres, fourteen quadrupolar spheres, fourteen quadrupolar spindle shapes, eight hexadecapolar octahedra and an instruction booklet. The models are fun to assemble and surprisingly stable but, inevitably, of somewhat limited application.

Models incorporating magnets embedded in styrofoam have been used to demonstrate $\pi$ overlap in olefins and *cis-trans* isomerism [17]. Two cylindrical pieces of styrofoam, each containing a suitably oriented magnet, are placed parallel between carbon centres with attached substituents. The pairs of magnets are attracted, so bringing the styrofoam cylinders together and forming the $\pi$ bond. The substituents may be arranged in either the *cis* or the *trans* orientation. If the magnets are arranged so as to repel rather than attract, further chemical features may be demonstrated such as rotational barriers in ethane or staggering of the cyclopentadienyl rings in ferrocene.

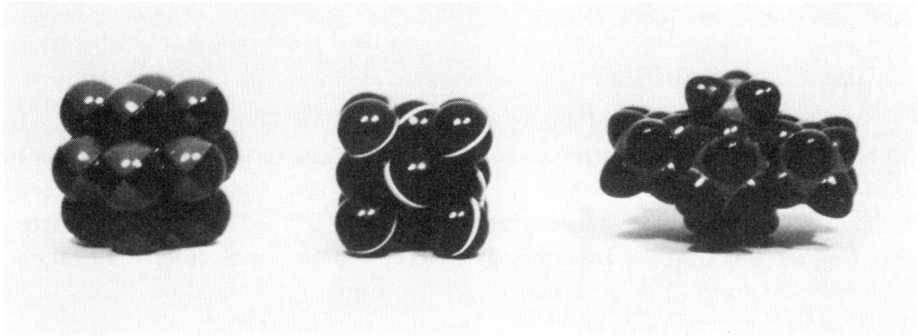

Fig. 8.1 – Kihara: molecular crystals of (a) hexamethylenetetramine (body-centred cubic), (b) carbon dioxide (face-centred cubic), (c) uranium hexachloride (hexagonal symmetry).

Two-dimensional models incorporating magnets can be used in conjunction with an overhead projector. It is possible to obtain strips of magnetic material with oppositely polarised faces. These strips may be attached to flat styrofoam models of water molecules, for example, with two similarly polarised strips on the parts of the model representing hydrogen and an oppositely polarised strip at the oxygen. If several of these are floated on water in a transparent dish and the dish is placed on an overhead projector stage, the 'hydrogen bonding' of the 'molecules' can be observed on a projection screen [18].

A kit of spheres with magnets embedded at one end and counter-balanced by lead at the other can be obtained from SRM. Four colours are used and when the container is agitated spheres come into contact and are attracted, simulating chemical processes in a simple way.

## 8.4 RAFTS AND TWO-DIMENSIONAL MODELS

Bubble rafts are a well-known means of showing close-packing of spheres in two dimensions. Bubbles can be formed on the surface of water containing a few drops of detergent simply by blowing in air through a fine capillary. 'Defects' and discontinuities occur naturally but can be induced by varying the pressure. Polystyrene spheres floating on water [19] are rather more permanent and 'defects' may be introduced by putting in spheres larger or smaller than those forming the raft. Interactions of flat, magnetised molecular shapes floating on water are described in 8.3 above [18].

The Atomix is a manufactured model consisting of two five inch square parallel acrylic plates between which is a shallow square well containing six thousand ball bearings 1 mm in diameter. The two dimensional array formed as a result can be arranged vertically or horizontally and can, therefore, be placed on the stage of an overhead projector. When it is vertical, the ball-bearings 'close-pack' in two dimensions. Discontinuities similar to grain boundaries and dislocations form and these may be moved by tapping the model lightly. After shaking, careful manipulation of the model will cause the balls to simulate the process of crystal growth. Phase changes and surface phenomena can also be demonstrated by shaking the model, because electrostatic charges acquired during the shaking process cause some of the balls to remain suspended as though they were, for example, vapour molecules above a liquid or solid surface.

The voltage supplied to the vibrator unit of the Molecular Dynamics Simulator can be controlled by a Variac. Because the motion of the spheres depends on the voltage, the Variac is therefore analogous to a temperature control. When the spheres are supported on a flat glass plate, kinetic behaviour of gases and liquids can be simulated and also condensation and crystallisation phenomena, among many others. Defects of various kinds can be demonstrated.

For elementary work, a much simpler model supplied by MLI consists of a tray containing a collection of coloured marbles which have to be agitated by hand. The same company also supplies a three-dimensional kinetic model where spheres vibrate in a perspex cylinder.

## 8.5 MISCELLANEOUS DYNAMIC MODELS

### 8.5.1 Manufactured

Both Griffin and George and MLI produce simple cubic lattice models in which the spheres are joined by very flexible springs. The models were designed to demonstrate the vibration of atoms or ions in solids. Instructions for use are provided with the Griffin and George model.

Two models from Leybold are of interest. One of these demonstrates the transformation of $\gamma$- to $\alpha$-iron in which the structure changes from face-centred cubic to body-centred cubic. The model comprises a metal base on which thirteen vertical spokes are mounted which support table tennis balls threaded on them. The transformation is effected by means of a lever which moves the rods and the vertical positions of the balls. A small ball representing carbon atom is also incorporated so that it can be demonstrated that this will only fit into the face-centred cubic structure without distortion. The model is therefore useful when discussing steel manufacture. The Leybold 'Model Glide Plane' consists of a model of part of a crystal lattice (three rows) which is linked to a tension and compression gauge so that stresses produced when a row of atoms is moved may be monitored. The effect of 'foreign' (larger) atoms can also be investigated. The name of this model is misleading but it is possible that the fault lies in an inaccurate translation from German.

### 8.5.2 Homemade

This brief Chapter ends with a few examples of widely differing models reported in the literature. The only properties which they have in common is that they are all models and all, according to the definition at the beginning of 8.1 (above), dynamic.

One of the Tested Overhead-Projection Series (TOPS) in the *Journal of Chemical Education* is a demonstration of the crystal structures of diamond and graphite [20]. A box with transparent sides has parallel grooves at the top and bottom into which transparent sheet plastic inserts can slide. Black dots painted on each insert show the positions of atoms in a particular crystal plane so that when they are all aligned the crystal structures of diamond and graphite may be visualised in three dimensions once the assembly is projected. When the inserts are moved, the graphite layers appear to slide over each other.

Carraher's polymer models [21] are made either from poppet beads joined together or from polystyrene spheres or corks threaded on string. The limited solubility range of polymers is demonstrated by adding the 'polymer chain' to a beaker of spheres representing solvent molecules and showing how relatively difficult it is to contact all of the chain with spheres on shaking. Similar experiments indicate other polymer properties including viscosity and excluded volumes of polymer molecules.

An unusual application of paper tetrahedra models is to use them to demonstrate the gyroscopic properties of water [22] both as vapour and liquid. Directions are given for preparing a model consisting of a ring of six tetrahedra, and this is intended to represent a low temperature condensed species in liquid water. A model of this kind can be suspended and rotated in such a way that precessional motion may be observed.

Another model of a pseudorotation process, this time to show all the possible conformers of cyclohexane, is described by Strauss [23]. Six spheres represent the ring atoms and these are attached to rods which are pivoted near the end away from each sphere so that the sphere can move vertically. The extreme end of each rod rides on a cam which determines the conformation of the spheres. More than one cam may be used. A chair form is made either by removing the cam or by using a flat cam and bending alternate spheres above and below the pivots. For boat, twist and intermediate forms a cam with two maxima and two minima round its edge is required. A small motor can be used to rotate the cam, when the twist and boat forms will alternate. Intermediate forms are produced by bending the rods as for the chair conformation and then rotating them. Manipulation of cam and spheres to give any conformation is possible, including a 'half-chair' in which one sphere is above the plane formed by the others.

Hartman [25] describes a model in which molecular motion is simulated by vibrating springs. It is motor driven and can be operated by the students themselves.

**Table 8.1**
Dynamic models

| Model | Price code (Table 1.4) |
|---|---|
| Walden inversion model (Berry and Botterill) | (b) |
| Pseudorotation model (Berry, Hedley and Hoppé) | (b) |
| Plastic Stereomodels | (b) |
| Kihara models | (b) |
| †SRM Chemical Reaction Demonstration Kit | (a) |
| †SRM Atomix | (b) |
| Molecular Dynamics Simulator | (d) |
| MLI 2-D Kinetic Model Kit | (b) |
| Griffin and George Atom Model (with springs) | (a) |
| MLI Atom Model (with springs) | (a) |
| Leybold Crystal Model of Steel Transformation | (b) |
| Leybold Model Glide Plane | (d) |

†These have not appeared in recent SRM catalogues.

## REFERENCES

[1] Berry, R. W. H., (a) *Educ. in Chem.*, **11**, 182 (1974); (b) *ibid.*, **14**, 102 (1977).
[2] Berry, R. W. H. and Botterill, C. J., *Educ. in Chem.*, **4**, 139 (1967).
[3] Berry, R. W. H. and Hoppé, J. I., *Educ. in Chem.*, **8**, 28 (1971).
[4] Berry, R. W. H., Hedley, R. and Hoppé, J. I., *Educ. in Chem.*, **11**, 94 (1974).

[5] Berry, R. W. H., *J. Chem. Phys.*, **32**, 933 (1960).
[6] Gootjes, J. and Bakuwel, G., *J. Chem. Educ.*, **42**, 407 (1965).
[7] Skvortsov, I. M., Jones, M. J. and Elvidge, J. A., *Chemy. Ind.*, 105 (1969).
[8] Petersen, Q. R., *J. Chem. Educ.*, **47**, 24 (1970).
[9] Jones, H. R. and Bentley, R. B., *Proc. Chem. Soc.*, 438 (1961).
[10] Jones, H. R., *Educ. in Chem.*, **2**, 25 (1965).
[11] Nyquist, H. L., *J. Chem. Educ.*, **42**, 103 (1965).
[12] Hamon, D. P. G., *J. Chem. Educ.*, **47**, 398 (1970).
[13] Riess, J. G., *J. Chem. Educ.*, **50**, 850 (1973).
[14] Broomhead, J. A., Dwyer, M. and Meller, A., *J. Chem Educ.*, **45**, 716 (1968).
[15] Levenson, R. A., *J. Chem. Educ.*, **52**, 386 (1975).
[16] Kihara, T., (a) *J. Phys. Soc. Japan*, **15**, 1920 (1960);
(b) *Acta Crystallogr.*, **16**, 1119 (1963);
(c) *Acta Crystallogr.*, **21**, 877 (1966).
[17] Meek, J. S., *J. Chem. Educ.*, **48**, 112 (1971).
[18] Barnard, W. R., *J. Chem. Educ.*, **45**, 341 (1968).
[19] Booth, N., *Educ. in Chem.*, **1**, 18 (1964).
[20] Alyea, H. N., *J. Chem. Educ.*, **45**, A225 (1968).
[21] Carraher, C. E., *J. Chem. Educ.*, **47**, 581 (1970).
[22] Morrow, F. J., *J. Chem. Educ.*, **48**, 368 (1971).
[23] Strauss, H. L., *J. Chem. Educ.*, **48**, 221 (1971).
[24] The use of these models was demonstrated by Professor Dreiding at the symposium on *Chemical Education in Europe* held in Madrid under the auspices of the Federation of European Chemical Societies, September 7th and 8th, 1975.
[25] Hartman, K., *J. Chem. Educ.*, **53**, 111 (1976).

# 9

# Construction devices and techniques

## 9.1 INTRODUCTION

Construction aids for the preparation of homemade models are only cursorily considered in previous Chapters on the assumption that it would be more helpful to the reader if the information were, in general, gathered under one heading rather than dispersed throughout the text. However, there are two exceptions to this in that both a template designed by Bassow [1] and the commercially-available Kristakit ball-punch are described in Chapter 4.5.1 (Figure 9.1). References [5]-[14] inclusive at the end of Chapter 4 also contain information on construction techniques, some of which are discussed below. Reference is made to devices used not as construction aids but for *measuring* distances and angles in skeletal models in Chapter 5.2.2 and 5.2.3.

Fig. 9.1 – Kristakit ball-punch

## 9.2 JIGS AND TEMPLATES

These assist in the construction of molecular or crystal structure models by holding the spheres in position while they are punched, drilled or cut in the appropriate places. Therefore for space-filling molecular models, a jig would be used to indicate the orientation and depth of the cuts required to remove segments from a sphere so as to produce a typical Stuart-type model. For space-filling ionic or ball-and-spoke models, the function of the jig is to mark positions or punch holes at the points where spheres are to be glued together or connectors inserted. Apart from Kristakit, few jigs appear to be available on the commercial market but two others are described here and several designs have been reported in the literature. These last are frequently combined efforts of technical and academic staff and are usually made up in the school or institution workshop if metal or wood is used.

The Griffin 'Atomic' Model Jig consists of an aluminium body with accurately drilled holes and two piercing tools with wooden handles. The body contains a removable annulus and so polystyrene spheres of three sizes (19, 25 and 38 mm) can be used. Because of the selection and range of angles available, all the common symmetries are possible with co-ordination numbers from two to six and twelve-co-ordination for both types of close-packing as well. Instructions for use are included.

SRM market a Molecular Model Jig which is exactly similar to the Griffin one. Both are extremely useful, durable tools and moderately priced.

The Griffin Crystal Lattice Jig is constructed in a similar way but there are two annuli and three piercing tools. While all the angles and symmetries of the Atomic Model Jig are available, others which permit construction of models of the crystal systems and all the Bravais lattices are also possible. Instructions are again provided with the jig. The cost is moderate and would soon be recouped by saving on the purchase of manufactured models.

There are basically two ways in which the preparation of a sphere with tetrahedrally disposed 'bonds' can be approached. A tetrahedral arrangement can be regarded either as two mutually perpendicular pairs of 'bonds' or as three points with 120° angles between and a fourth perpendicular to the plane of the first three. These views are reflected in the techniques used. For many common symmetries it is necessary to mark several points on the sphere in one plane and others in a plane at right angles to it so a number of jigs are built on the 'two circle' principle. A relatively early version originated in the crystallography laboratories of the University of Cambridge [2] and one of its descendants was used by Greenwood and Smith [3] to prepare models of borane structures. The wooden balls used already had holes bored through their centres so that they could be threaded on a spigot attached to a protractor. Another protractor was placed at right angles to this one so that the ball could be rotated in either of two mutually perpendicular planes, and held in a particular orientation, which was accurately known, while it was drilled. With a device of this kind any drilling symmetry is possible and it could be readily adapted to take cork spheres for the preparation of space-filling molecular models. In this case segments perpendicular to the drilling direction could be cut off instead of drilling. Nowadays polystyrene spheres could be used rather than cork for this purpose.

Ballard, Goffen and Price [4] employ the two protractor principle in their jig. The drill can be inserted through a drill guide, which rotates in a vertical plane so that the degree of rotation can be measured by means of a protractor. Perpendicular to this is another protractor which can be used to measure the extent of rotation in a horizontal plane. The sphere can be locked in position while drilling takes place and then moved on to the next accurately known drilling position. Although the jig was designed to locate 'bond' positions and to drill holes for connectors for ball-and-spoke models, it can also be adapted for the preparation of Stuart models. The size of sphere needs to be different and, instead of a drill, a rotating spoke-shave can remove segments perpendicular to the bond direction. Polystyrene spheres can also be used but they can be penetrated by a hot wire passed through the drill guide instead of being drilled.

A jig in which the polystyrene sphere rests in a wedge-shaped hole or trough, with the angle between the sides 70°, has been designed for preparing space-filling atom models with tetrahedrally disposed faces [5]. A gap exists between the bottom of the sloping wedge and a vertical wooden stop against which the sphere also rests. The size of this gap determines the depth of the segment to be cut off because the cut is made vertically level with the edge of the wedge. The cut face is then put against the side of the wedge and a second cut made. The third and fourth cuts are made with two faces in contact with the wedge walls. The device is held in a support which facilitates accurate cutting. An alternative is to mount the trough horizontally but to arrange sawing or cutting slots at 70°-71° [6] (71° quoted). The internal angle should be 70.5° (180°-109.5°) but it is unlikely that this degree of precision could be achieved. The sphere is allowed to drop a fixed amount through a hole in the trough until it rests against the stop. The distance between the edge of the trough and the stop, which can be adjusted with spacers if required, determines the depth of cut. After the first cut, positions for subsequent cuts are determined as before.

Marking of 'bond' locations in spheres for ball-and-spoke models can be accomplished in a number of ways apart from those already described. A tetrahedral arrangement can be marked out [7] if a hemispherical hole of the same diameter as the sphere to be marked is made in a wooden block and then a pin is knocked through from the outside so that it protrudes through the centre of the hole. Three lines 120° apart are then marked on the top of the block. Next, three triangular needle guides of grooved plywood are made so that when these are stuck on to the lines radiating from the centre of the block, the top edge of each makes an angle of 19.5° with the horizontal. If the sphere is placed on the pin, one bond site is marked and the sphere is in position for further marking. This is effected by running a thin knitting needle down each of the needle guides so that it penetrates to the centre of the sphere each time. Hence the angle between any two holes in the sphere is 109.5°.

Another possibility is to make a hole in the block which is only 93% of the diameter of the sphere [8]. When the sphere is placed in it (assuming the depth of the block is sufficient) the edge of the hole marks a circle joining three corners of the tetrahedron so that marking at the point where three lines at 120° impinge on the sphere is sufficient. The fourth corner is at the top of the sphere. Many polystyrene spheres have pole 'pips' as a result of the moulding process, in which case one pole 'pip' can be placed on the pin while the opposite one gives the fourth position. However, with this method care has to be taken over the angle of entry of the spokes when joining spheres. With the first method this is already correctly determined.

A different technique for punching tetrahedrally disposed holes in spheres in order to construct ball-and-spoke assemblies is used by Conard and Bent [9]. Two vertical brass rods guide a slide carrying a flat V-shape, with the point uppermost. The V can move up or down, rather like a guillotine (revolutionary not stationery!). Another V-shape is fixed to the base, point down, with its arms at right angles to those of the upper V. The internal angle in both cases is 70.5°. The V shapes are made of oak and holes are drilled perpendicular to the sides and through the arms. A wooden dowel punch, sharpened at the end, can then pass through these. When the sphere is in position, it rests on the lower V and the upper one is brought down on to it. Punches pushed down through the appropriate holes in the Vs then penetrate the sphere producing tetrahedrally arranged 'bonding' sites into which wooden dowel spokes can be inserted.

The distance between 'bonding' sites on the surface of a sphere can be determined by multiplying the bond angle in radians by the radius [10]. A strip of paper cut to this length can be fixed on to a sphere with pins. To find a third site at the same angle, compasses set to the length of the piece of paper are used to draw arcs from each site already located. The third point is where these arcs intersect. A similar use of compasses but also including marking off segments to be removed when preparing space-filling atom models is described by Bassow [1].

Simple templates of stiff paper, card, hardboard or wood are easy to make and information is available from a number of sources on their construction and use [1], [5], [8], [11], [13]. Essentially they are made by drawing a series of lines at various angles which intersect at a common centre on a square or rectangular piece of the chosen material. The angles most frequently required are 109.5°, 90° and 120°. A circle with the same diameter as the sphere to be used is then drawn about the centre and cut out. If paper is used it is best first to glue it on to card or hardboard so as to strengthen it before cutting out the circle. It can then be placed over the top of an empty box with a depth greater than the radius of the sphere.

A slight rim, marking the equator, can frequently be observed on polystyrene spheres. Both this and the pole 'pips' already mentioned make useful guides when inserting the sphere into the hole for punching, because the equator and hole edge should coincide. Pencil, felt-tip pen or compass point can be used to mark the sphere, while punching can be accomplished by using a corkborer pin or a suitably sharpened metal rod such as welding rod. The punch must be used at the correct angle and should penetrate to the centre of the sphere. One way of ensuring this is to use either a double piece of corrugated cardboard [1] or two similar pieces of hardboard bolted together, [5], [12] because the punches can be inserted *between* the two layers along the marked guidelines, which can be grooved in the case of hardboard by drilling. If handled with care such templates can be used to punch tetrahedral, octahedral or cubic arrangements of holes for the construction of three-dimensional arrays. However, it should be pointed out that any errors in 'bond' location are rapidly magnified as the structure grows and initial practice with models of small molecules is recommended. A printed paper protractor template which can be stuck on to card or plywood is available from Griffin and George for use with their 'Polyzote' spheres, but it can be used with any polystyrene spheres of suitable size (from 0.5 to 2.0 inch diameter in six stages). An instruction book giving construction details is supplied.

All the above examples are concerned with marking or cutting spheres. A device used by Jagannathan [15] describes jigs which may be employed in the construction of skeletal model units. For example, two wooden triangular prisms, each with a tetrahedral angle and grooved on the outer edges forming this angle, may be placed at right angles to each other. The grooves are then tetrahedrally disposed. Plastic rod or tube may then be placed in the grooves and joined by glueing at the centre. A square block of wood with two grooves at right angles cut in it and a hole bored through the centre can be used to make octahedral units. A similar block with three grooves at 60° can be used for the construction of trigonal bipyramidal units. The units can be joined by stiff tape.

## 9.3 ADHESIVES

Although ordinary carpenter's glue takes a long time to dry and sometimes flakes off so-called 'vinyl cements', PVA and proprietary brands such as Evostick and Bostick join virtually any materials likely to be used in models. A satisfactory glue can be made by dissolving pieces of polystyrene in toluene (**not** benzene, which is too toxic) but it is best to make the glue using high density expanded polystyrene because the low density material is so insubstantial that much labour is involved in getting enough of it into solution. The final viscosity should be about that of treacle. A glass rod can be used to dispense the glue. An electric glue gun provides the least messy means of dispensing exactly the right amount of glue for joining polystyrene spheres or permanently fixing connectors in place, but this might be deemed to be something of a luxury unless very extensive model-building were contemplated.

I have recently found an answer to a problem which has concerned me for some time, that of finding a 'temporary' adhesive for polystyrene spheres so that they could be taken apart and used again. A material which is suitable in every respect for this purpose is 'Blu-Tack'. The spheres are held together for as long as is required by placing a small piece of 'Blu-Tack' on one of them and pressing them together. When dismantling is necessary, they are simply pulled apart and the 'Blu-Tack' can also be pulled off and used again. It is a cheap, readily obtainable material which is clean and easy to handle.

## 9.4 HOT WIRE CUTTERS

These are useful devices for shaping polystyrene, particularly for the preparation of space-filling atom models where a neat accurately placed cut is important. However, it should be emphasised that the fumes produced when polystyrene is pyrolysed are toxic. No-one should use a hot wire cutter on polystyrene for a prolonged period and the operation should be performed either in a well-ventilated room or with the apparatus in a fumecupboard. The hot wire melts its way through the polystyrene so that the surface produced is much smoother and neater than could be obtained with a knife or saw. The procedure is also considerably quicker because the piece to be cut need only be held up against the wire and pushed slowly forward until the cut is complete. Awkward shapes such as orbital lobe models are made more easily than by any other method apart from moulding. As there are no moving parts the apparatus is safer to handle than most cutting devices, providing due regard is given to the fume problem.

The commercial models vary in degree of sophistication and type of accessory. For example, the Griffin model consists of a table on which to place the work through which a vertically supported taut wire passes to a clamping device. A sculpturing tool is also supplied. A control box contains two heating units, one for the wire and one for the sculpturing tool, and instructions for use are included in the kit. The Pyro/Kerf model also has two heating intensities and is based on the same design principle, though accessories include a foot switch and a protractor set (Figure 9.2). By means of the protractor, T-square and guide edges, all of which can be clamped to the table, the piece of polystyrene to be cut can be placed accurately in position and retained in the correct alignment with respect to the hot wire during the cutting procedure. Neither sculpturing tool nor instructions are supplied with this equipment and the accessories are not included in the cost of the hot wire cutter itself.

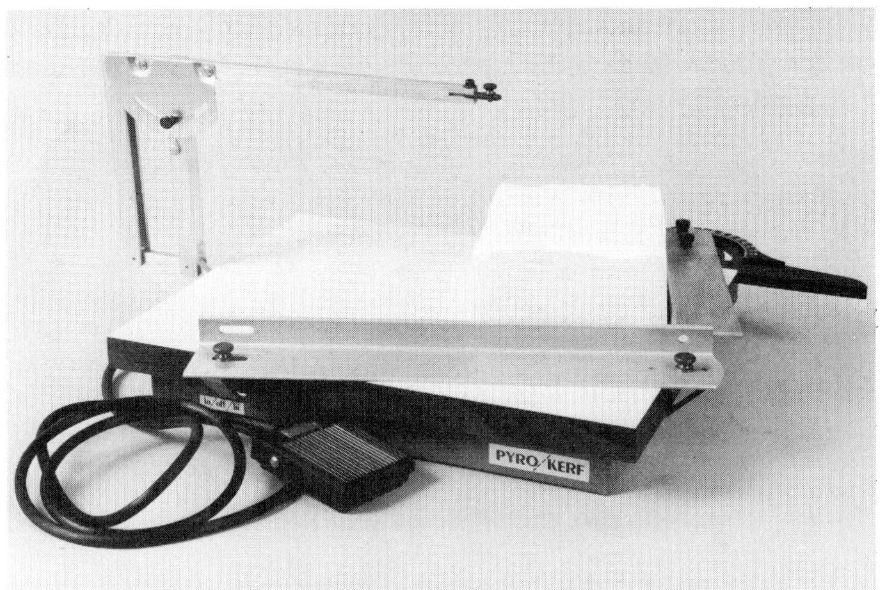


Fig. 9.2 – Pyro-kerf hot wire cutter.

A homemade hot wire cutter in which retorts and clamps support the base and the vertical taut wire is described by Pritchard [13]. Nichrome wire (SWG26) carrying a current of two amps is used. A rheostat may be included in the circuit in order to vary the current if required. The plywood base has a slit cut three-quarters of the way through the middle. This acts as a guideline for the heated wire. Another piece of wood is arranged vertically and acts as both a 'back-stop' and a support for the polystyrene. This can move backwards or forwards by means of slots in the base. A variant on this is mentioned by Ormerod [14]. He recommends a 180 mm length of SWG30 nichrome wire which would carry approximately a two amp current when it is connected to a six volt car battery or transformer. A rheostat may be included if desired. This device can be used in shaping atomic and molecular orbital models.

**Table 9.1**

Construction devices

| Device | Price code (Table 1.4) |
|---|---|
| Kristakit | b |
| Electric glue gun | b |
| Griffin hot wire cutter | c |
| ‡Pyro/Kerf hot wire curter | c |

‡A simpler version is now available from SRM; price code 'b'.

## REFERENCES

[1] Bassow, H., *Construction and Use of Atomic and Molecular Models*, Pergamon (1968).
[2] Megaw, H. D., *Brit. J. Appl. Phys.*, **4**, 107 (1953).
[3] Greenwood, N. N. and Smith, J., *Educ. in Chem.*, **1**, 25 (1964).
[4] Ballard, R. E., Goffen, D. B. and Price, N., *Lab. Practice*, **18**, 1189 (1969).
[5] Tetlow, K. S., *Educ. in Chem.*, **1**, 7 (1964).
[6] Walker, R. A. and Davis, L. F., *J. Chem. Educ.*, **42**, 417 (1965).
[7] Bentley, S. C., *School Sci. Rev.*, **48**, 488 (1967).
[8] Sanderson, R. T., *Teaching Chemistry with Models*, Van Nostrand (1962).
[9] Conard, C. R. and Bent, H. E., *J. Chem. Educ.*, **46**, 492 (1969).
[10] Spargo, P. E., *J. Chem. Educ.*, **51**, 259 (1974).
[11] Dabby, R. E., *Making Crystal Models*, Pergamon (1969).
[12] Yeoman, G. D., personal communication. (Modified version of the Tetlow jig (reference [5]) prepared and used by student chemistry teachers in the Education Department, Nottingham University).
[13] Pritchard, D. M., *School Sci. Rev.*, **53**, 139 (1971).
[14] Ormerod, M. B., *The Architecture and Properties of Matter*, Edward Arnold (1970).
[15] Jagannathan, K., *J. Chem. Educ.*, **53**, 47 (1976).

# 10

# Two-dimensional models; pictures, projections and computers

## 10.1 INTRODUCTION

This Chapter is chiefly concerned with images of models. These may range from simple drawings on paper or blackboard to three-dimensional stereographic projeçtions. The former are not considered in detail here though examples are given of some exceptional photographs, drawings or paintings of models which are of outstanding artistic or scientific merit.

Other topics include the use of overhead or slide projectors and computer methods.

## 10.2 TWO-DIMENSIONAL MODELS AND PRINTED IMAGES

The authors of a book [1] entitled *Chemistry: its Role in Society* are clearly convinced of the value of models as teaching aids because the book incorporates cardboard molecular models which can be punched out and used to illustrate the text. Another type of planar model ('Magimolecular') produced by Magiboard will adhere to any clean polished surface such as glass or a gloss-painted wall (Figure 10.1). The shapes can be removed and rearranged, or stored and used later. The colour-coded plastic shapes representing atoms are like planar Stuart models (Chapter 2.2) and molecular models are assembled by placing 'bonding' edges (as opposed to faces in the 3-D models) in contact. The scale is 20 mm ≡ 0.1 nm. A number of simple molecules can be portrayed including the optical isomers of lactic acid, for example. Of course, ionic compounds cannot be shown even approximately correctly, although the instruction sheet does suggest representing sodium chloride by placing two discs of suitable dimensions side by side! Though planar models of this type may be better than drawings of structural formulae in some circumstances, they are not adequate substitutes for 3-D models. The great advantage of the latter is that they can be handled and examined from various view points, and nowadays a small molecular models kit could be purchased for about the same price as the planar models. However, a 3-D kit can usually only be used by one student at a time while the planar models are suitable for classroom demonstrations. Moreover, a 'Magimolecular' stencil enables the same shapes to be drawn on paper or transparent sheets for use with an overhead projector.

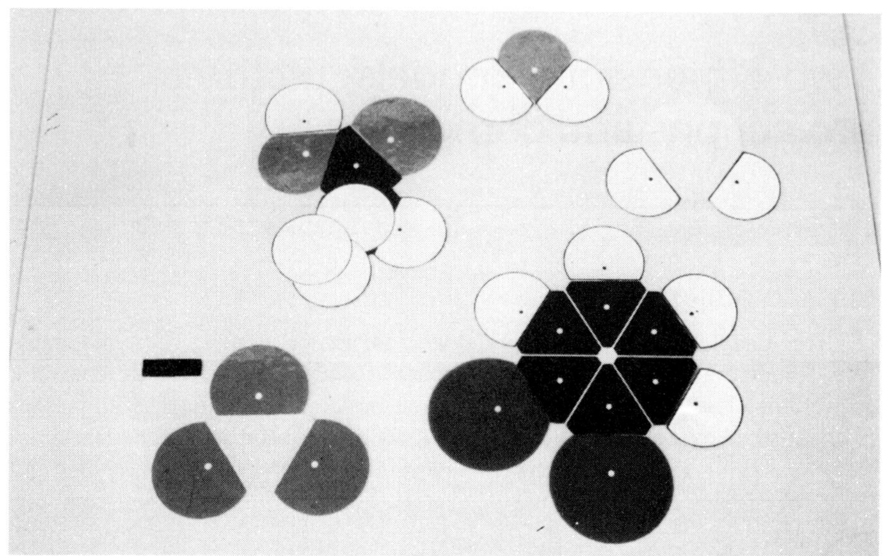

Fig. 10.1 – Some two-dimensional 'Magimolecular' models.

Planar models made from coloured poker chips, cardboard or paper are described by Cavagnol and Barnett [2]. Glue-backed magnetic tape is stuck on to the models and they can then be made to adhere to a metal 'blackboard'. They can be used to illustrate the structure of atoms and molecules and chemical reactions. Dissociation of ions in water is demonstrated using models of water molecules made from coloured paper backed by card and poker chips for the ions. The models are also used to portray the principles of osmosis and enzyme-substrate interactions.

Franke's book *Sinnbild der Chemie* [3] reproduces a unique collection of photographs, drawings and oil paintings derived from models representing all the important categories described in this book, as well as some others. The photographs are of superb quality and in some cases variation in depth of focus has been used to give a three-dimensional effect. Colour is used in eleven of the fifty-six plates and in one of the thirteen drawings. Study of this book is both aesthetically and scientifically rewarding. A similar work is Pauling and Hayward's *The Architecture of Molecules* [4] which has fifty-seven colour plates of drawings of a variety of model types. The pictures of the extended frameworks of the diamond and Prussian Blue crystals are particularly striking. Paintings of models reproduced in Phillip's paper on enzyme structures [5] are also of interest because these coloured pictures show how the substrate fits into a cleft in the enzyme molecule.

Nevertheless, no photographs, paintings or drawings, no matter how carefully produced (see for example the computer-based perspective drawings of orbitals by Jorgensen and Salem [6]), are truly three-dimensional images. Apart from projection methods (see below), the only technique known to me which results in a genuine three-dimensional effect is the Xograph process. Essentially, this effect is obtained by dividing the picture into a very large number of parallel vertical strips. Embossed in register with this is a plastic coating by means of which some images are blocked out and others enhanced. The viewer's eyes see different images simultaneously. This interesting process has been used successfully to photograph CPK space-filling molecular models (Chapter 2.3.2) of lysozyme and its substrate [7]. The 3-D colour photographs appear alongside the text and they are remarkably successful though somewhat garish.

## 10.3 MODELS AND THE OVERHEAD PROJECTOR

The existence of octahedral holes in close-packing sequences can be demonstrated by stacking two layers of close-packed spheres on the stage of an overhead projector. The holes show up as bright spots on the screen. When the octahedral holes are filled by smaller spheres, these spots disappear. Specific features of a skeletal molecular model can be emphasised by changing the focus when the model is placed on the projector stage. For example, it is possible to show whether a terminal hydrogen is axial or equatorial by the sharpness of its image relative to the rest of the molecule. This is useful when only one model is available for a class of students and also when movement of the model might change the conformation.

Bubble rafts produced on the surface of a transparent liquid in a transparent container placed on the overhead projector stage can also be projected on to a screen. Hollow transparent spheres can be used as well, and then it is possible to introduce 'defects' into the two-dimensional close-packed layer by the addition of spheres differing in size. Magnetic simulation of electrostatic phenomena using projected shadows of models is described by Barnard [8]. For example, strips of appropriately polarised magnetic material can be attached at suitable sites on two-dimensional polystyrene models of water molecules. When these are allowed to float on water the attractions and repulsions cause them to simulate the hydrogen bonding process.

An analogue method for calculating atomic co-ordinates and dipole moments using Dreiding molecular models (Chapter 5.2.2) has been devised by McEachern and Lehmann [9]. A shadow projection technique is employed. The model is placed in an open transparent box with three mutually perpendicular sides. An overhead projector provides the light source. Projections are obtained by successively lining up the sides of the box (containing the model) perpendicular to the light beam. The positions of atoms or ends of bonds are recorded on paper taped to the box side which receives the image using a back projection technique. The dipole moment can be calculated using these two-dimensional components.

## 10.4 USE OF FILM AND SLIDE PROJECTORS

A film entitled 'Molecular Models and Substitution Reactions' illustrates $S_N^1$ and $S_N^2$ reactions by means of molecular models and projection formulae. Film mixing techniques and animated cartoons are used. The film is in colour and runs for just under 23 min. Further details are given in Reference 22.

An extensive range of colour slides of inorganic crystal structure models is available from Polyhedral Solids. The catalogue lists over two hundred and includes structures based on both kinds of close-packings, 'mixed' packings, polyhedra, structures containing trigonal prisms and a series of over fifty depicting 'relationships', for example zinc blende and wurtzite.

Slides of models projected using the time-lapse technique have been used to demonstrate both the equivalence of octahedral axes and the pinacol-pinacolone rearrangement [10]. In the former case, colour transparencies of the model with different coloured axes in different orientations were obtained. With the time-lapse process, one image fades as the next appears and they are partially superimposed. In this way the relationship between the various orientations of the model could be understood more readily. A sequence of twenty slides of ball-and-spoke models was used to demonstrate the pinacol-pinacolone rearrangement, again using time-lapse projection and so giving the impression that the model was moving. As a result the reaction appeared to be happening on the screen. In their article on lap-dissolve projection, Harpp and Snyder [11] mention the use of the technique in portraying an $S_N^2$ reaction by means of photographs of molecular models.

On the other hand, McGrew and Kitzman [12] filmed a moving model. The model was placed on a rotating stand and rotated about an axis which was the perpendicular bisector of the film image. Ball-and-spoke models were used. Suitable adjustment of the focal length of the camera and the rotational period of the model resulted in a stereoscopic effect, students reporting 'moderate to strong 3-D impressions'. Three films illustrating concepts in conformational analysis were made and the student response seems to have been favourable.

Undoubtedly the next best thing to the model itself is a really good 3-D projection of it. Perhaps the most useful application of 3-D slides is to *remind* students of models which they have already examined, although it is sometimes necessary to use the 3-D image as a substitute for the model, particularly when insufficient time or money is available for the assembly or purchase of a large range of models. In such cases 3-D slides are most helpful and slides of excellent quality can now be produced. While we are concerned primarily with projection techniques in this Section, it is important to note that stereoslides can be used for individual viewing [13], [14] and, alternatively, that stereoprints can also be made [14]. These are particularly valuable because they can be reproduced in books and journals and so reach a much wider public than slides.

The preparation of stereoslides is discussed in detail by Smith, [14] who favours a stereocamera (one with two lenses 70 mm apart) chiefly because a stereo mounting service is available for the slides produced while with a single lens camera the operator has to do his own mounting. However, Smith describes a commercially available slide bar which is used to move the camera the requisite amount between the two shots and he also gives directions for making a slide bar from simple materials. On the other hand, McGrew prefers a single lens camera, mainly because the stereocamera has a fixed lens separation which he considers to be too wide for close-up photography of models [13]. He describes a bracket on which the camera is mounted which permits adjustment of the interlens distance to any required value. An alternative to McGrew's own design is a commercially available rack-and-pinion unit which is, inevitably, much more expensive.

A neat method of mounting the model is used by McGrew. It is supported on a Plexiglas plate bolted to a rod and clamp arrangement which does not appear in the photographs. The background is of poster board, supported vertically by a pegboard panel. The background is slightly out of focus in the photograph, and with correct adjustments of interlens spacing and distance the model appears to be suspended in space. Ideally, the front of the model should appear to project from the screen, the middle to lie in the plane of the screen and the back to lie behind it.

The chief difficulty seems to be in mounting the slides and misalignment in the slide holders will prevent accurate superposition of the two images. However, a simple illuminated viewer can be used to assist in the alignment of slide pairs [15].

A stereoprojector can be used to project the stereopair slides. Once each image is focussed they are brought together so that overlap is virtually complete. Polarising filters are placed in front of the slides while the audience wears polaroid glasses and sees a 3-D image on the screen. The screen itself must not be depolarising, so it is usually covered with aluminium paint or foil. Aerosol silver spray can be used for the finish [14]. Alternatively, duralumin sheet mechanically buffed to a satin finish is satisfactory [15]. Because use of the polarising filters cuts down overall light intensity, the slides should tend to be over- rather than under-exposed. A powerful light beam is also necessary. Suitable backgrounds are red or yellow, but black has proved satisfactory for Dreiding models. Instead of using a stereoprojector, it is possible to use an arrangement which consists of two projectors coupled together so that their operations synchronise [15], [16]. With such a set-up, stereoprojection, ordinary slides and time-lapse projection can be used together in the same programme.

Stereoprojection has been successfully applied to the production of 3-D images of models of macromolecules [14], for example Biobits (Chapter 6.2.2.i). Stereoslides of ball-and-spoke and skeletal models can be used effectively to demonstrate various stereochemical phenomena including isomerism and conformational analysis in inorganic chemistry [13], [15], [16] and the geometry of extended arrays representing inorganic crystal structures as well [17].

A short note by Gelbard [18] describes a simple technique for making inexpensive stereoscopic slides by photographing a molecular model from a 'right eye view' and then a 'left eye view'. The resulting pair of slides should be observed through 'two separate eye viewers'.

## 10.5 MODELLING BY COMPUTER

Computers can be used to display the conformational behaviour of molecules in a very elegant manner [19]. Although attempts have been made to predict conformations of molecules by calculating the free energies of all possible conformations using a computer, in the case of a complex molecule such as a protein a number of simplifying assumptions have to be made [20] and the reliability of the process is therefore reduced. Whether or not the computer is programmed to produce a model of recognisably the same form as the three-dimensional ones with which we are familiar, it is still necessarily a model of some kind. Frequently a rather stylised picture of a ball-and-spoke or skeletal model is used. Alternatively, distribution of electron density in a molecule may be portrayed. In the hands of a skilful programmer, the computer can be made to simulate many molecular properties very satisfactorily, particularly those involving motion. In his television series *The Ascent of Man*, Bronowski used computer-graphics to demonstrate various processes, among them the stacking of the base pairs in the DNA molecule. Here, in Bronowski's own words, 'The gene is forming visibly in front of our eyes'. The computer-graphic technique was also used to demonstrate the rotation of the DNA molecule in such a way that the observer appeared to be able to view the molecule from every possible angle [21]. In this way an impression of the three-dimensional quality of the molecule itself may be gained.

Over three centuries lie between Pepys' model of a ship and computer-graphics of the DNA molecule but the essential features of models remain unchanged. The description 'pleasant and useful' may still appropriately be applied to many of today's molecular and crystal structure models.

## REFERENCES

[1] Chickos, J. S., Garin, D. L. and Rouse, R. A., *Chemistry: Its Role in Society*, D. C. Heath (1973).

[2] Cavagnol, R. M. and Barnett, T., *J. Chem. Educ.*, **53**, 643 (1976).

[3] Franke, H. W., *Sinnbild der Chemie,* Basilius Press, Basel (1966).
[4] Pauling, L. and Hayward, R., *The Architecture of Molecules,* Freeman (1964).
[5] Phillips, D. C., *Sci. Amer.,* **215,** No. 5, 78 (1966).
[6] Jorgensen, W. L. and Salem, L., *The Organic Chemist's Book of Orbitals,* Academic Press (1973).
[7] Harte, R. A., and Rupley, J. A., *J. Biol. Chem.,* **243,** 1663 (1968).
[8] Barnard, W. R., *J. Chem Educ.,* **45,** 341 (1968).
[9] McEachern, D. M. and Lehmann, P. A., *J. Chem. Educ.,* **47,** 389 (1970).
[10] Hubinger, H. and Schultz, H. P., *J. Chem. Educ.,* **48,** 618 (1971).
[11] Harpp, D. N. and Snyder, J. P., *J. Chem. Educ.,* **54,** 68 (1977).
[12] McGrew, L. A. and Kitzman, K., *J. Chem. Educ.,* **50,** 407 (1973).
[13] McGrew, L. A., *J. Chem. Educ.,* **48,** 531 (1971).
[14] Smith, I., *RIC Reviews,* **4,** No. 1, 19 (1971).
[15] Buckingham, J., *Educ. in Chem.,* **11,** 159 (1974).
[16] McGrew, L. A., *J. Chem. Educ.,* **49,** 195 (1972).
[17] Buckingham, J., personal communication.
[18] Gelbard, G., *J. Chem. Educ.,* **53,** 792 (1976).
[19] Ollis, W. D., *Proc. Roy. Inst.,* **45,** 1 (1973).
[20] Levinthal, C., *Sci. Amer.,* **214,** 42 (1966).
[21] Bronowski, J., *The Ascent of Man,* BBC (1973).
[22] Presented by Suckling, C. J., produced by Thomson, G.; 16 mm, sound, colour, 23 min., 1976. Available from Dept. of Pure and Applied Chemistry, University of Strathclyde.

# Appendix I

Manufacturers and Suppliers of Models

| Name and Address | Model Supplied |
|---|---|
| Ace Plastic Co. Inc.,<br>91-30 Van Wyck Expressway,<br>Jamaica N.Y. 11435<br>U.S.A. | Plastics spheres and acrylic rods |
| Ace Scientific Supply Co. Inc.,<br>1420 E. Linden Avenue,<br>Linden N.J. 07036,<br>U.S.A. | Agents for Büchi and RJM Exports |
| Addison-Wesley Publishing Company,<br>West End House,<br>11 Hills Place,<br>London W1R 2LR | Disc and HGS Maruzen |
| S. Andersson<br>Sölvegatan 8B<br>223 62 Lund<br>Sweden | Polyhedral Solids and slides of models |
| Becton Dickinson U.K. Ltd.,<br>York House,<br>Empire Way,<br>Wembley,<br>Middlesex HA9 0PS | Agents for Schwarz-Mann (CPK models) |

| Name and Address | Model Supplied |
| --- | --- |
| Beevers Miniature Models,<br>Simon Square Centre,<br>Pleasance,<br>Edinburgh 8,<br>EH8 9HW | Beevers |
| W. A. Benjamin Inc.,<br>2 Park Avenue,<br>New York, N.Y. 10016<br>U.S.A. | Disc and HGS Maruzen. |
| Bronwill Scientific Inc.,<br>Box 7824-Lyell Sta.,<br>11·28 Lexington Avenue,<br>Rochester N.Y. 14606,<br>U.S.A. | Godfrey space-filling and skeletal |
| Büchi Laboratoriums-Technik AG,<br>CH–9230 Flawil,<br>Switzerland. | Dreiding, including 'Plastic Dreiding'; also agents for Rinco |
| Cambrian Chemicals Ltd.,<br>Suffolk House,<br>George Street,<br>Croydon, CR9 3QL | Agents for Science Related Materials (SRM) |
| Capital Biotechnic Developments Ltd.,<br>66a Churchfield Road,<br>London, W3 6DL | Atomunits and Biobits |
| Catalin Ltd.,<br>Waltham Abbey,<br>Essex.<br>(see Needs Plastics Ltd., below) | Catalin covalent and Catalin ionic |
| Central Scientific Co.,<br>Division of Cenco Instruments Corp.,<br>2600 S. Kostner Avenue,<br>Chicago, Il1.60623,<br>U.S.A.<br>and in Europe – | Cenco-Petersen. |

| Name and Address | Model Supplied |
|---|---|
| Cenco Instrumenten B.V.,<br>P.O. Box 3336,<br>Konijnenberg 40,<br>Breda,<br>The Netherlands. | |
| Cochranes of Oxford Ltd.,<br>Fairspear House,<br>Leafield,<br>Oxford, OX8 5NT. | Minit, Orbit and Unit |
| Crystal Structures Ltd.,<br>Bottisham<br>Cambridge, CB5 9EA | Ball-and-spoke assemblies, kits and components; Geodestix, Norbury orbitals, Effex Stereo |
| Dyna-Slide Co. – see Science Related Materials (below) | |
| The Ealing Corporation,<br>Science Teaching Division,<br>225 N. Massachusetts Avenue,<br>Cambridge, Mass. 02140,<br>U.S.A.<br>and in Europe – | Atomunits, Biobits, CPK, Kendrew and Kennard-Dorě, Scale Atoms, some SRM. |
| Ealing Beck Ltd.,<br>15 Greycaine Road,<br>Watford, WD2 4PW | |
| Edmund Scientific Co.,<br>555 Edscorp Building,<br>Barrington N.J. 08007,<br>U.S.A. | Space-filling, spheres, 'Think-sticks', polyhedra* |
| Elford Plastics Ltd. –<br>see Philip Harris, below. | |

*Only 'Think-sticks' in recent catalogue.

| Name and Address | Model Supplied |
|---|---|
| Esselte Studium<br>S-11285 Stockholm,<br>Scheelegatan 24,<br>Sweden. | Scale Atoms |
| Estrutural,<br>Rua Almeida E. Sousa<br>14-1.ºE.-Lisboa-3,<br>Portugal. | Spheres; assembled and built on base |
| Fisher Scientific Co.,<br>Stansi Educational Division,<br>711 Forbes Avenue,<br>Pittsburgh, Pa. 15219,<br>U.S.A. | F.H.T. and Scale Atoms. |
| and in Europe –<br>Fisher Scientific Co.,<br>Zeltweg 67,<br>8032 Zürich,<br>Switzerland. | |
| and<br>Fison Scientific Apparatus Ltd.,<br>Bishops Meadow Road,<br>Loughborough,<br>Leics. | F.H.T. |
| A. Gallenkamp and Co. Ltd.,<br>P.O. Box 290,<br>Technico House,<br>Christopher Street,<br>London, EC2P 2ER. | Linnell and PEEL – now marketed by Griffin and George – see below. |
| Griffin and George Ltd.,<br>285, Ealing Road,<br>Alperton,<br>Wembley, Middlesex,<br>HA0 1HJ | Courtauld, polystyrene spheres, ball-and-spoke, Chemical Structures Frame, d-orbital model, Geoframe, polyhedra, various kits, hot wire cutter, moulded and foam plastics, jigs. |

| Name and Address | Model Supplied |
|---|---|
| Philip Harris Ltd.,<br>Lynn Lane,<br>Shenstone,<br>Staffordshire,<br>WS14 0EE | Polystyrene spheres, Scale Atoms, Molymod, Minit, Orbit and Unit. |
| Heinemann Educational Books Ltd.,<br>48 Charles Street,<br>London, W1X 8AH. | Scale Atoms. |
| Herzbi Ltd.,<br>37a–39 Leswin Road,<br>London, N16 7NX. | Addatoms. |
| Holt, Rinehart and Winston Inc.,<br>383 Madison Avenue,<br>New York, N.Y. 10017,<br>U.S.A.<br><br>and in Europe –<br>Holt-Blond Ltd.,<br>120 Golden Lane,<br>London, EC1Y 0TU. | Molecular Dynamics Simulator. |
| V. A. Howe and Co. Ltd.,<br>88 Peterborough Road,<br>London, SW6. | F.H.T. |
| Incentive Lärosystem,<br>P.O. Box 474,<br>S-17104 Solna 4,<br>Sweden. | Scale Atoms. |
| Industrial and Scientific Instrument Co.,<br>5225 Germantown Avenue,<br>Philadelphia, Pa. 19144,<br>U.S.A. | Space-filling, ball-and-spoke, polyhedra, orbitals. |

| Name and Address | Model Supplied |
|---|---|
| Klinger Scientific Apparatus Corp.,<br>83–45 Parsons Boulevard,<br>Jamaica, N.Y. 11432,<br>U.S.A. | Agents for CSL (see above) and Leybold (see below). |
| Koch-Light Laboratories Ltd.,<br>Colnbrook,<br>Bucks., SL3 0BZ. | Dreiding, including 'Plastic Dreiding'. |
| Lab-Aids Inc.,<br>130 Wilbur Place,<br>Bohemia,<br>New York 11716,<br>U.S.A. | Skeletal kits. |
| Labquip,<br>Ashridgewood Place,<br>Forest Road,<br>Wokingham,<br>Reading, RG11 5RA. | Nicholson |
| LaPine Scientific Co.,<br>6001 S. Knox Avenue,<br>Chicago, Ill. 60629,<br>U.S.A. | Spheres-in-contact, ball-and-spoke, polyhedra and orbitals. Also Leybold models (see below). |
| G. O. Larson,<br>Route 2, Box 234,<br>Big Rapids,<br>Michigan,<br>U.S.A. | Cubelts. |
| LeMont Scientific Inc.,<br>Pike Street,<br>LeMont, Pa. 16851,<br>U.S.A. | Cork and plastics spheres and spheres-in-contact assemblies. |

| Name and Address | Model Supplied |
|---|---|
| Leybold Heraeus GMBH and Co. KG,<br>5-Köln-51,<br>Bonner Strasse 504,<br>W. Germany. | Leybold space-filling molecular and crystal structure, ball-and-spoke and dynamic. |
| and in the U.K.<br>Leybold-Heraeus Ltd.,<br>173 Greenwich High Road,<br>London, SE10 8JA. | |
| Macalaster Scientific Company,<br>Box R. Dept. CSM,<br>Nashua,<br>New Hampshire 03060,<br>U.S.A. | Spheres, ball-and-spoke, polyhedra, DNA. |
| Macmillan Education Ltd.,<br>Hound Mills,<br>Basingstoke,<br>Hants. | Orbit |
| Magi-White Display Boards Ltd.,<br>153 Clapham High Street,<br>London, SW4. | Magimolecular. |
| Maruzen Company Ltd.,<br>Foreign Book Department,<br>6-Tori-Nichome,<br>Nihonbashi,<br>Tokyo 103,<br>Japan. | HGS/Maruzen. |
| Mitamura Riken Kogyo Inc.,<br>2-3 Hongo Bunkyo-ku,<br>Tokyo,<br>Japan. | Kihara |
| Morris Laboratory Instruments Ltd.,<br>480 Bath Road, Slough,<br>Bucks. SL1 6BH. | Kristakit. |

| Name and Address | Model Supplied |
|---|---|
| Needs Plastics Ltd.,<br>15 Queensway,<br>Enfield,<br>Middx. EN3 4SG. | Catalin covalent and ionic |
| Orme Scientific Ltd.,<br>P.O. Box 3,<br>Stakehill Industrial Estate,<br>Middleton,<br>Manchester. | Agents for Büchi (see above) and Rinco (see below). |
| Oxford University Press,<br>Ely House,<br>37 Dover Street,<br>London, W1X 4AH. | Wells |
| Plasteel Corporation,<br>26970 Princeton,<br>Inkster,<br>Michigan 48141,<br>U.S.A. | Variety of Polystyrene shapes, coloured pipe-cleaners. |
| Polyhedral Solids,<br>Sölvegatan 8B,<br>22362 Lund,<br>Sweden. | Polyhedra and slides. |
| Prentice-Hall Inc.,<br>Englewood Cliffs,<br>N.J. 07623,<br>U.S.A.<br><br>and in the U.K.<br>Prentice-Hall International,<br>66 Wood End Lane,<br>Hemel Hempstead,<br>Herts. HP2 4RG. | FMM and PMDM. |

| Name and Address | Model Supplied |
|---|---|
| Rinco Instrument Co.,<br>503 S. Prairie Street,<br>Greenville, I11. 62246,<br>U.S.A. | Dreiding and Fieser. |
| R.J.M. Exports Ltd.,<br>Leafield,<br>Oxford. OX8 5NT. | Minit, Orbit and Unit<br>cf. Cochrane. |
| Sargent-Welch Scientific Co.,<br>7300 N. Linder Avenue,<br>Stokie, Ill. 60076,<br>U.S.A. | Godfrey space-filling, polyurethane spheres and ball-and-spoke kit. |
| S.A.S.M.,<br>99 Rue Oberkampf,<br>Paris XI$^e$,<br>France. | Space-filling, spheres, ball-and-spoke, orbitals and other SASM models. |
| Schwarz-Mann,<br>Mountain View Avenue,<br>Orangeburg,<br>N.Y. 10962,<br>U.S.A. | C.P.K. |
| Scientific Glass Apparatus Co. Inc.,<br>735 Broad Street,<br>Bloomfield, N. J. 07003,<br>U.S.A. | Dreiding, F.H.T., Godfrey. |
| Science Labs.,<br>Box 737,<br>El Cerrito,<br>California 94530,<br>U.S.A. | Various space-filling, spherical, ball-and-spoke, skeletal. |
| Science Related Materials Inc.,<br>P.O. Box 1422,<br>Janesville, Wis. 53545,<br>U.S.A. | SRM models in all major categories. Also market Minit and Orbit. |

| Name and Address | Model Supplied |
|---|---|
| Spiring Enterprises Ltd.,<br>North Holmwood,<br>Dorking,<br>Surrey. | Molymod. |
| Stansi Educational Division,<br>Fisher Scientific Co. Ltd.,<br>P.O. Box 1099,<br>Montreal, Quebec,<br>Canada. | FHT and Scale Atoms – see under Fisher for U.S.A. and European addresses. |
| United Technical Supplies Ltd.,<br>29 Tottenham Court Road,<br>London, W.1. | Kihara. |
| VWR Scientific International,<br>P.O. Box 3200,<br>San Francisco,<br>California 94119,<br>U.S.A. | Godfrey. |
| John Wiley and Sons Ltd.,<br>Baffins Lane,<br>Chichester,<br>Sussex PO19 1UD. | Minit and Orbit. |
| The Windmill Press,<br>Kingswood,<br>Tadworth,<br>Surrey. | Scale Atoms. |
| S. Wylie (Manufacturing) Co. Ltd.,<br>149–151 Edinburgh Avenue,<br>Trading Estate, Slough,<br>Bucks. | Walden inversion and pseudorotation models. |

# Appendix II

Named Models and Their Suppliers

| Model | Manufacturer and/or Supplier |
|---|---|
| Addatoms | Herzbi |
| Atomix | SRM |
| Atomunits | Capital Biotechnic |
| Biobits | Capital Biotechnic |
| Catalin | Needs Plastics |
| Cenco-Petersen | Central Scientific |
| Courtauld | Griffin and George |
| CPK | Schwarz-Mann, Becton Dickinson, Ealing |
| Cubelts | Larson |
| Disc | Addison-Wesley, W.A. Benjamin |
| Dreiding | Büchi, Koch-Light, Rinco |
| Fieser | Rinco |
| FHT | Fisher, Fisons, Howe, Stansi |
| FMM | Prentice-Hall |
| Geodestix | CSL |
| Geoframe | Griffin and George |
| Godfrey | Bronwill, Sargent-Welch, VWR Scientific |
| HGS/Maruzen | Addison-Wesley, Benjamin |
| Kendrew | Ealing |
| Kennard-Doré | Ealing |
| Kihara | Mitamura Riken Kogyo, United Technical |
| Kristakit | Morris Laboratory Instruments |
| Minit | RJM Exports/Cochrane, Wiley, Philip Harris, SRM in U.S.A. |
| Molecular Dynamics Simulator | Holt, Rinehart and Winston |

| Model | Manufacturer and/or Supplier |
|---|---|
| Molecular Fragments | SRM, Cambrian in U.K. |
| Molymod | Spiring Enterprises, Philip Harris |
| Nicholson | Labquip |
| Norbury Orbitals | CSL, Klinger |
| Orbit | RJM Exports/Cochrane, Macmillan, Philip Harris, Wiley, SRM in U.S.A. |
| PEEL | Gallenkamp/Griffin and George |
| PMDM | Prentice-Hall |
| Polyhedral Solids | S. Andersson |
| Pyro-Kerf hot wire cutter | SRM/Cambrian |
| Scale Atoms | Esselte Studium/Incentive Lärosystem, Ealing, Fisher, Heinemann, Philip Harris, Windmill Press |
| Unit | RJM Exports/Cochrane, Philip Harris |
| Wells | Oxford University Press |

Note. Models known by the name of their manufacturer or supplier are not listed here.

# Index

# Notes

## Notes

# Notes

## Notes

# Notes

# Notes

## Notes

The importance of modelling throughout science is implicit in its employment to demonstrate project feasibility: and in chemistry as an aid to elucidate structures of crystals and molecules. This book, by a specialist with an international reputation, surveys molecular and crystal structure models currently available to scientists in universitities, polytechnics and schools for teaching at all levels, and to industrial and academic laboratories for research. No other book known to the publishers gives such an up-to-date and detailed account, or such wide coverage. The author discusses theoretical background and practical use, which is justified in terms of the expanding interest in structural chemistry and the effectiveness of models as teaching aids. This relates to the modernisation of chemistry teaching everywhere by demonstrating the structural principles at the earliest stage.

The book provides a theoretical and practical guide to all users, discussing the whole range of models available for use today, from the variety of sophisticated manufactured varieties to the simple homemade devices which can be prepared by individuals from readily available and inexpensive materials. The discussion of employment of models in research is a unique feature of special interest to industrial and other research workers.

**Readership:** Workers in chemistry, crystallography, mineralogy, geology, geochemistry, molecular biology, medicine and metallurgy in teaching, research and industry.